The Awakening

Does scientific evidence support the
existence of a divine Creator?

C. Robert Follett

"The Awakening" by C. Robert Follett.
ISBN 978-1-60264-166-2 (softcover);
ISBN 978-1-62137-027-7 (electronic copy).

Library of Congress Control Number: 2012908675.

Doreen

In Loving Memory

To my children

Dawn, Chip, Rob, Marni and Kevin

Through developing wisdom

we can achieve "awakening".

- Jamin Sunim

Contents

Charts, Tables and Diagrams

Acknowledgements

Chapter 12

CBS News "60 Minutes" – Steve Kroft interview with Dr. J. Craig Ventor

Chapter 14

Dr. Ed Neeland, University of British Columbia – Opinion on Evolution

Preface

The battle lines were drawn: the scientific world waging war against the religious status quo, and religion waging war against science.

In one encampment: the scientific community, with doctorates from cutting-edge universities and access to the latest technological advances in programming and equipment, are wielding weapons of proven scientific facts, as well as yet-to-be-proven theories. The Hubble telescope is only one of many very sophisticated tools in their arsenal. Launched in 1990, its gaze has been responsible for solving many of the mysteries of our universe, including its age.

Supercomputers using modeling programs allow scientists to project into the future, as well as into the past, in order to predict outcomes with extreme accuracy. Scientists have dreamed of replicating natural life for many years. Recently, in the field of biology, living matter has finally been synthetically created. It has no previous DNA.

In the other encampment: the religious community, equipped only with an ancient book written thousands of years ago by unknown writers. Their only weapon is faith.

Actually, this war has been raging for many years, beginning with Darwin's Theory of Evolution, and intensifying in the last few decades. Today you can even see evidence of this struggle on our highways. One vehicle may be sporting the stylized Christian symbol of a fish on its rear bumper, while the next vehicle may display that same fish symbol with the name Darwin included in the design.

Evolution and the origin of species has been taught in some public schools as early as 1925. When the Russians launched Sputnik, the first satellite into space, in 1957, many became convinced that the U.S. was falling behind in science—a concern that initiated the funding of the *Biological Sciences Study* by *The National Science Foundation*. This study was influential in returning evolution to high school biology textbooks. By the 1960s, evolution was widely taught, even though it was still only a theory. And yet, the teaching of creationism—the belief that the universe and mankind were created by God—has been legally challenged on every front. Today, the only place creationism may be taught is in private schools and in home schooling situations.

Each new scientific discovery seems to chip away at the credibility of the Bible-thumping sermons on Sunday mornings, in steepled buildings and on the televisions of many of America's homes. How will God's word possibly be able to survive under such crippling attacks from every side? Many of today's most brilliant scientists, during their childhood, themselves spent Sunday mornings sitting in church pews and praying to a God that they now completely disavow. What could they have possibly learned that could turn their beliefs around by 180 degrees?

My search has led me to some exciting conclusions. I believe these findings to be compelling evidence of a mysteriously unique concord between science and religion, but not until the layers of false beliefs and opinions are peeled away can the truth be revealed. After years of standing at my personal fork in the road, of sleepless nights and diligent searching, I can offer you the enlightenment of a fresh, new perspective on the age-old dilemma . . . what to believe. "The Awakening" has come as a result of my personal journey, and I wish now to share it with you.

Good luck,

The author

Chapter 1

The Core Issues

Famed psychiatrist and TV host Dr. Phil McGraw has a favorite expression: "No matter how flat you make a pancake, it still has two sides."

This is equally true in the war between science and religion. There are two core issues in this dispute:

1.   How and when did the universe begin?

2.   What are the origins of modern man?

Surprisingly, both sides do agree on one thing: that the Universe and modern man both had a beginning. Aside from that, there is little else they can agree on.

Science on the universe: Science believes the universe began with a huge explosion—the Big Bang—that blew particles of gas and dust into space, which later cooled to form the various galaxies, quasars, and ultimately the solar systems.

Science does not see evidence of intelligent design, nor does it have any explainable reason for this occurrence.

Religion on the universe: Religion believes that a divine God was the designer and creator of the universe, for the purpose of providing for mankind's existence.

I believe it is quite difficult for anyone to fully comprehend time in billions of years when comparing events in relation to one another. At the end of this chapter, and throughout this book, you will find timeline charts. Leaps in scale will be made to better illustrate various events.

Science on the origin of man: Science believes that life began with a single-cell organism, which originated in the waters of Earth after billions of years, when the conditions became ideal for supporting life. They believe that life ignited itself when the proper elements came together by chance, and then evolved into more complex living structures through a process known as natural selection. This is where a mutation or shift occurs over time, which protects a species from predators or certain conditions that may otherwise eliminate their species. A good example of this is the katydid, which became camouflaged to look like a leaf.

Science believes these original cell structures, through evolution, first became Chordates (eel-like), then Tetrapods (fish), then Reptiles (lizard), then Mammals (rodent), then Primates (monkeys), then Homididae (apes), then Homo (Man's direct ancestor).

Science does not see evidence of intelligent design, nor does it have any explainable reason for this occurrence.

Religion on the origin of man: Religion believes that God created man in His own image.

Before we set out on this search for answers, let us remember that seldom is anything truly black or white. Normally, answers come in various shades of gray.

For instance, I realize that there are some scientists who believe in God, just as there are those in religion who believe in evolution. So, when I make declarative statements regarding beliefs, I'm making those statements on the basis of a general consensus, rather than one of all-inclusiveness.

As we take this journey together, even though there may be some difficult terrain to maneuver, I am certain that our quest for answers will result in your achieving new understanding of what have proven to be some of life's oldest and most perplexing mysteries.

Did you know. . .

. . . that the Bible does not address the question "How old is the Universe?"

. . . that there is nothing biblical supporting a 6,000-year-old earth?

Chart #1

**14.5 billion year timeline of Universe
(Universe's estimated age)**

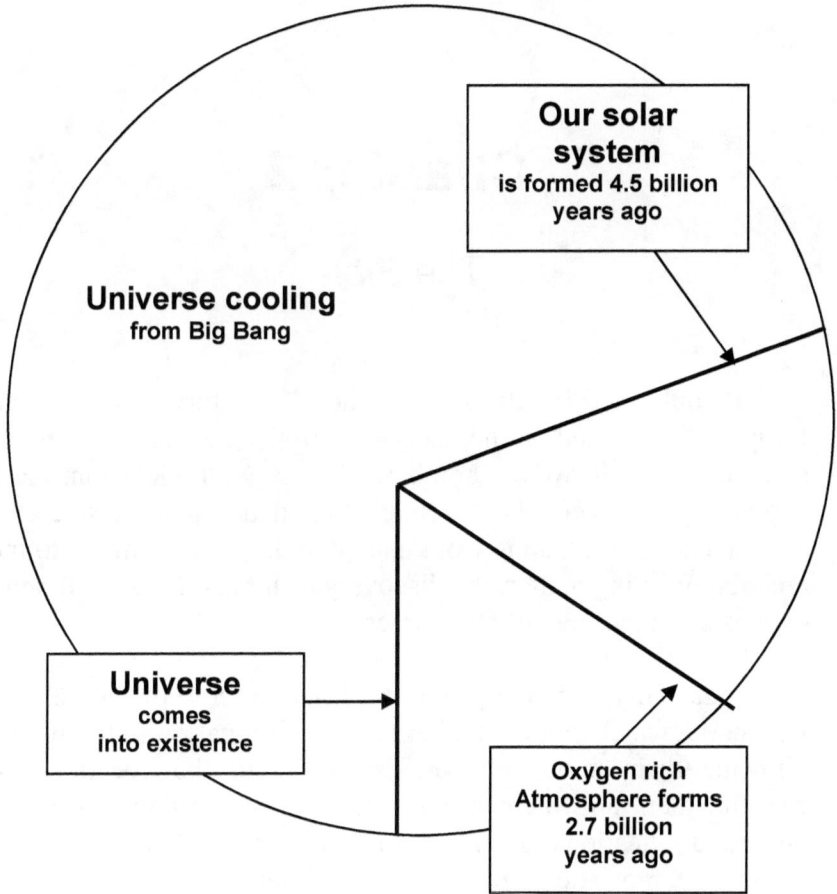

**Our solar
system**
**is formed 4.5 billion
years ago**

Universe cooling
from Big Bang

Universe
**comes
into existence**

**Oxygen rich
Atmosphere forms
2.7 billion
years ago**

Past ←——→ **Present**

Chapter 2

The Bible

It would be difficult to live in the 21st century without being fairly well informed on the many accomplishments in the field of science. After all, we are bombarded daily with television, radio and newspaper accounts of remarkable advances in the sciences. We have seen everything from man's landing on the moon to the creation of living matter, the discovery of human DNA, and rapid expansion in the field of electronics.

Today, many people carry the latest version of the cellular telephone, which can guide them to any destination in the world, allowing them to take pictures or video of that location and transmit them anywhere in the world, as they enjoy a menu of endless applications and listen to their favorite tunes, all while reading a famous novel on their smart phone.

But there remains a cloak of mystery around man's oldest source of knowledge—the Holy Bible, religion's only tool. I think we would be remiss if we did not arm ourselves with some basic knowledge about mankind's oldest and most distributed publication, at 30 million annually, before continuing on our journey. I contend that even many who do read their Bible on a daily basis are clueless to the many truths waiting to be unlocked—if they only knew how to interpret the words.

What is the Bible?

The Bible is the inspired word of God, which was written down by man. I realize that this may be a difficult concept for many people to wrap their minds around, especially if they were not reared in a faith-based family environment.

But since you're obviously still reading this book, you must be a person with some genuine curiosity about the subject of God. If you will only keep an open mind through the end of this chapter, I believe you'll begin to see that there is something credible about my claims.

The Holy Bible of today contains a total of 66 books—39 in the Old Testament and 27 in the New Testament. The Old Testament was written in Hebrew and Aramaic, the New Testament in Greek. Both were translated into English in 1380 A.D.

There are fourteen books, called Apocryphal books, which are not included in most Bibles because the Jews did not consider them to be inspired scripture, nor were they ever quoted by either Christ or his Apostles. However, at the Council of Trent, in 1546 A.D., the Catholic Church accepted six of these books as inspired and added them to their Bible.

Many things attest to the fact that the Bible truly is the word of God. Most important among them are the 333 prophecies in the Old Testament, which were fulfilled hundreds of years later in the New Testament during the life of Christ. I believe that only God could have made such predictions with one hundred percent accuracy.

What method did God use to communicate with man?

God has spoken directly to man: **Matthew 3:17** - And lo a voice from heaven, saying, This is my beloved Son, in whom I am well pleased.

God has spoken through an angel: **Hebrews 2:2-3** - For if the word spoken by angels was stedfast, and every transgression and disobedience received a just recompence of reward; 3 How shall we escape, if we neglect so great salvation; which at the first began to be spoken by the Lord, and was confirmed unto us by them that heard him;

God has spoken through prophets: **Acts 3:21** - Whom the heaven must receive until the times of restitution of all things, which God hath spoken by the mouth of all his holy prophets since the world began.

God has spoken through Jesus: **Hebrews 1:1-2** - God, who at sundry times and in divers manners spake in time past unto the fathers by the prophets, [2]Hath in these last days spoken unto us by his Son, whom he hath appointed heir of all things, by whom also he made the worlds;

God has spoken through the apostles: **Acts 1:2** - Until the day in which he was taken up, after that he through the Holy Ghost had given commandments unto the apostles whom he had chosen:

God has spoken through visions: **Ezekiel 1:1** - Now it came to pass in the thirtieth year, in the fourth month, in the fifth day of the month, as I was among the captives by the river of Chebar, that the heavens were opened, and I saw visions of God.

God has spoken through dreams: **Daniel 2:1** And in the second year of the reign of Nebuchadnezzar, Nebuchadnezzar dreamed dreams, wherewith his spirit was troubled, and his sleep brake from him.

God has spoken through inspiration: **Timothy 3:16** - All scripture is given by inspiration of God, and is profitable for doctrine, for reproof, for correction, for instruction in righteousness:

God has spoken through revelation: **Galatians 1:15-16** - But when it pleased God, who separated me from my mother's womb, and called me by his grace,

[16]To reveal his Son in me, that I might preach him among the heathen; immediately I conferred not with flesh and blood:

God has spoken through a combination of methods: **Matthew 1:20** - But while he thought on these things, behold, the angel of the LORD appeared unto him in a dream, saying, Joseph, thou son of David, fear not to take unto thee Mary thy wife: for that which is conceived in her is of the Holy Ghost.

To clarify, inspiration (from the word inspire) inspires or influences man to come up with an answer on his own, while in the case of revelation (from the word reveal), God reveals the answer. Malachi, the last Old Testament prophet, predicted the next messenger that God would send to Israel would be Jesus Christ. And then there was a "silent period" of approximately 400 years between the two testaments.

While the Bible records the most reliable account of mankind over the ages, there are always those who will challenge its authenticity, pointing to some Bible story with claims that a city, prophet, king or kingdom has never been found to exist, never been acknowledged, accepted as true or ruled to be correct, at least not initially. But over the centuries, the Bible has never been proven wrong.

There have been many times that doubters have wagged their fingers in disbelief at claims made in the Bible, only to have those temporary victories dashed at a later date. While the Bible has ultimately been proven right on every issue, some scholars must initially challenge God's word, only to be proven wrong later. For centuries, man questioned, even doubted the existence of an ancient city named Selah, which was mentioned in the Old Testament. Allow me to offer the following story as a possible solution to the whereabouts of the missing city.

A Tale of Two Cities

Part I

There is an ancient, mystical city located in the mountains of Edom called **Petra**, which is never mentioned anywhere in the Bible. If you saw the movie *Indiana Jones and the Temple of Doom,* you have also seen Petra, where it was filmed. Along with its beautiful temples, shrines and dwellings carved into the face of tall cliffs, there were also caves, which served as burial grounds for its kings.

Petra was the capital city of the Nabataeans, an Arabian nomadic people who had captured the city from the Edomites in 300 B.C. During the time of Christ, Petra was surrounded on all sides by the Roman Empire. Strange that it was able to resist annexation by the mighty armies of Rome, but there were circumstances that existed, which helped it remain as an independent nation.

Aside from the fact that the city was difficult to approach because of its hostile terrain and high location, the Romans depended on the Nabataeans to act as guides to navigate their armies across the desert. The Nabataean people were mostly wealthy merchants who had well-established trade routes to the Far East. Over nearly three hundred years, they had built a sophisticated system of hand-dug cisterns through the desert along their trade routes. These cisterns held rainwater, but were totally hidden from view. Only the Nabataeans knew where they were located, so the Romans had little they could do but pay them for their services to act as desert guides for their armies and trade caravans—a nice "cash crop" for these desert nomads. Of course, their primary source of wealth was the frankincense, myrrh, and other spices and goods they brought back from the Far East.

This worked well for them until the Romans developed sea routes to the Far East, which circumvented their need for crossing the desert. The Romans then defeated the Nabataeans in 106 AD, and the region became part of the Roman province of Arabia. Initially, the name of the city was Rekumu, but after the Roman occupation it became Hellenistic (of Greek influence) and known as Petra.

Following two earthquakes and invasions by the armies of the Muslims in the fourth century A.D., and the Crusaders in the twelfth century A.D., knowledge of Petra's location became all but lost to the western civilization until early in the nineteenth century, when the Swiss explorer Johann Burckhardt found the ruins of Petra by tricking the Bedouin locals to take him there. Today, Petra is the latest must-see destination for western tourists.

Part II

There was an ancient city in the region of Mt. Seir, called **Selah.** It was occupied by cave-dwelling Horites.

Genesis 14:6 - And the Horites in their mount Seir, unto Elparan, which is by the wilderness.

The Horites were driven out by the Edomites.
Deuteronomy 2:12 -The Horims also dwelt in Seir beforetime; but the children of Esau succeeded them, when they had destroyed them from before them, and dwelt in their stead; as Israel did unto the land of his possession, which the LORD gave unto them.

The city became a stronghold for the Edomites—until Amaziah, king of Judah, brought an army of 300,000 to retake the area in 794-3 B.C. After a great battle in the Valley of Salt, Amaziah hurled 10,000 Edomites to their deaths from the cliffs of Selah.

2 Chronicles 25:11-12 - And Amaziah strengthened himself, and led forth his people, and went to the valley of salt, and smote of the children of Seir ten thousand.

[12]And other ten thousand left alive did the children of Judah carry away captive, and brought them unto the top of the rock, and cast them down from the top of the rock, that they all were broken in pieces.

The ancient Roman historian, Josephus, writes of this incident in his Antiquities: "...whom he brought to the great rock which is in Arabia, and threw them down from it headlong."

The Edomites later re-took the city of Selah.

Here's the rest of the story:

The Nabataeans captured their city from the Edomites in 300 B.C.

Both Selah and Petra are located in the region of Mt. Seir.

Both cities have high cliffs.

The Horites of Selah were cave dwellers. Petra has caves.

In the Semitic language, Selah means" rock, cliff, fissure."

In the Nabataean language, rekumu means rock.

In Greek, Petra means rock.

I believe it is quite obvious that the ancient cities of Selah and Petra are one and the same city, divided by hundreds of years. This should help confirm the fact that the Bible may be confusing at times, but it is never wrong when properly interpreted.

How do you believe?

If you were to stand on the average street corner in America and ask passersby these questions, you would likely get one of three responses:

	Person A	Person B	Person C
How long has modern man been in existence?	Thousands of years	Millions of years	I don't know
How old is the earth?	Thousands of years old	Billions of years old	I don't know
How old is the universe?	Thousands of years old	Billions of years old	I don't know

Person **A** was probably brought up reading and being taught to believe the literal translation of the Bible, and also that it was a sin to question God or God's word, so they just keep their head down and plow ahead. They may realize that their answer doesn't seem very plausible, but it's what they were taught, so they just keep the faith.

Person **B** believes in evolution. That's not to say that this person doesn't believe in the Bible or doesn't go to church regularly. This person just thinks that evolution makes more sense, and he/she really doesn't understand what the Bible is saying, anyway.

Person **C** is simply confused. There are so many conflicting arguments going around that they just quit having an opinion about it.

Can you see yourself in one of these people? Don't you have legitimate questions about God, man and the universe?

How to read the Bible

When reading the Bible, every statement must always be taken as being literal whenever it is totally clear that it was intended to be taken literally. Otherwise, it may be taken as figurative. But remember, a figure of speech never does away with the literal truth. A word of caution, however: the Old Testament was written for early man, not 21st century man, so on occasion you may need to—please forgive this worn-out cliché—think outside the box.

Are you tired of trying to find answers in the Bible, only to become more confused with all the apparent contradictions? Believe me, I've been there ahead of you, but I think you will find believable answers to the difficult questions you've always wondered about in the next chapter.

Did you know . . .

. . . that people before Noah's flood lived over 10 times longer than we do today? There is a logical reason for this and it is fully explained in Chapter 7.

Chapter 3

God, Man and the Universe

Genesis 1:1 - In the beginning, God created the Heaven and the earth.

This unforgettable Bible verse from the Old Testament, beautiful in its simplicity, is the opening verse in the first of five books by Moses, which make up the Pentateuch, still the Jewish Bible of today.

The book of Genesis is God's account of His dealings with mankind, from the beginning of time to the period when the Hebrews were in bondage in Egypt. This was just before the birth of Moses. You may recall the story of Moses in which the pharaoh, fearing that the Jewish population was exceeding that of the Egyptians, ordered all the Hebrew babies killed. Moses' mother, fearing for her baby's life, set him adrift in a reed basket in the Nile. The pharaoh's daughter found him and reared him as her own child.

The truly exceptional fact about the book of Genesis, the first book in the Old Testament, is that it was written entirely in retrospect. Since Moses did not exist yet during the period of time covered by Genesis, God had to have told Moses what to write through conversations, visions, dreams, inspirations and revelations. There was no other way he would have known God's dealings with man before he was even born.

Astronomers know that the universe is approximately fourteen and a half billion years old, and that our solar system formed about ten billion years after the big bang. That would make earth about four and a half billion years old. But in Genesis 1:2 – 2:3, God tells us he created the heaven and the earth in six days and rested on the seventh day.

Man has always been confused about the story of creation as told in the book of Genesis. I have heard many people question the Bible's version of creation by stating, "It wouldn't be possible for God to create heaven and earth in only six days." And I would have to agree. But *if* astronomers believe in a creator, I'm sure they would tell you that actually, He ignited the universe in about six seconds. The actual creating part took a few billion years longer.

See the timeline at the end of this chapter, which will help you visualize the age of the Earth in relation to the dinosaurs, which were in existence millions of years ago.

So why is God painting such an impossible scenario for man to believe? The answer is simple. God wasn't talking to you and me. He was talking to Moses.

Just take a moment and Imagine yourself standing in Moses' sandals, surveying all of your surroundings and contemplating God's work. You look out at the horizon in all directions and see that, with exception of a few hills, the earth is totally flat. Every morning you see the sunrise in the east, travel across the sky, and set in the west. And when the sun goes down, the moon also rises in the east and sets in the west, just like the sun. And all the stars in the heaven do the same thing. And while everything in the heavens moves around you, the earth stands still. It's clear that Moses would have naturally thought he was standing at the center of the universe, that the Earth was the origination point for everything, even Heaven (firmament). Notice that God discusses the creation of the Earth in Genesis 1:1-2 before he discusses the creation of heaven in Genesis 1:7-8.

Genesis 1:7-8- [7]And God made the firmament, and divided the waters, which were under the firmament from the waters, which were above the firmament: and it was so. [8]And God called the firmament Heaven. And the evening and the morning were the second day.

That would naturally make the Earth more important than anything else in the universe, which is just the way Moses would have seen it. It should be obvious that God wrote the creation story to fit Moses' viewpoint. There is no way that God could have possibly explained the complexities of the universe, nor should he have needed to. So the story of creation, as told in Genesis, was given to Moses in the only way God could have given it to him. All too often, modern man withholds his faith in God simply because he doesn't understand God's storytelling motives.

Let's revisit this point in a modern-day setting. You're home from work early, spending some quality time with your three-year-old son, tossing a ball in the back yard. Suddenly your son points to the sun and asks, "What's that, Daddy?" And you respond, "That's the sun." And like any three-year-old, he follows up with, "Why, Daddy?"

So you try for a simple explanation. "Well, honey, the sun comes up in the morning, so it's bright and you can see when you play. And it goes down at night, so that it's dark and you can sleep."
Parents have little exchanges like this on a daily basis, and it's totally harmless—but the fact is, you were being anything but honest. The sun never goes up or down. It stays exactly where it is. So, let's replay this story with total honesty.

Take it from "Why, Daddy?":

"Well, you see, son, actually you and Daddy are really tiny, standing on a giant ball that's floating way up in the sky. The giant ball is called Earth. You and Daddy are so tiny that the ball looks flat to us. And the reason we don't fall off the earth is because of a force called gravity. It works like a big magnet to hold us on. And the earth goes in a path around the sun. The path is called an orbit. And the sun is so big, and we're so far away from the sun, that it takes a whole year to orbit the sun.

"And while the earth is going around the sun, it's spinning, so that during the day we're facing the sun, so it's bright, and during the night we're facing away from the sun and it's dark. Now, while the earth is spinning around the sun, it's also wobbling slowly back and forth. When it wobbles one way, the sun's rays hit the earth where we are more directly. That's summer. And when it wobbles the other way, we're not in the sun's rays as directly, so it's colder. That's winter. Do you understand?"

You do realize, don't you, that your three-year-old would probably never ask you another "why" question?

My point is this. Even though you weren't being totally honest with your son in your initial story, in the broader sense, you had told him everything that a three-year-old needs to know. The same is true with God's story to Moses

Did you know . . .

. . . that God and his angels have physical bodies like you and me, and that they also eat regular food?

God and two of his angels stopped by Abraham's tent on the plains of Mamre and ate lunch with Abraham, on their way to assessing the moral conditions in the cities of Sodom and

Gomorrah. This is when God told Abraham that his descendants would outnumber the stars in the heavens, and that Abraham's 80-year-old wife, Sarah, would conceive and give him an heir. (See Genesis 18:1-33)

Notice that Abraham recognized God immediately, which would indicate that they had met previously on other occasions.

Chart #2
Earth's 4.5 billion year timeline
(Earth's estimated age)

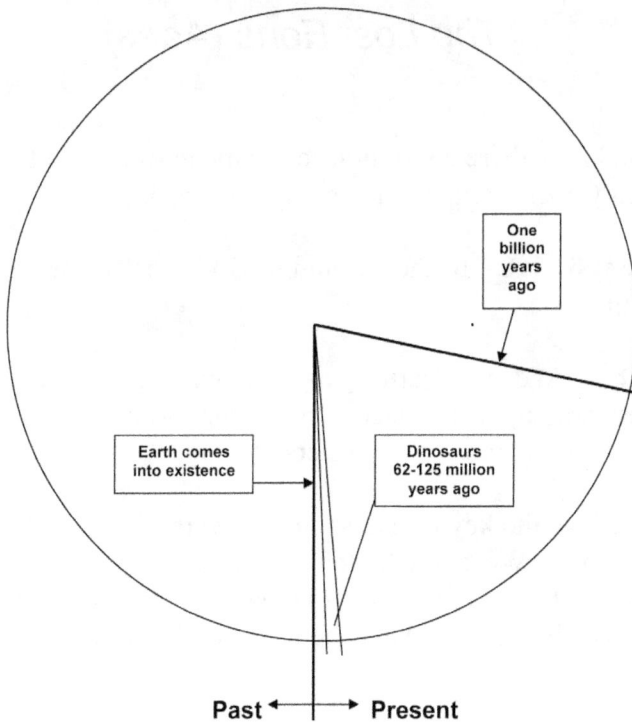

One
billion
years
ago

Earth comes
into existence

Dinosaurs
62-125 million
years ago

Past ←┼→ Present

Chapter 4

The Lost Eons (Ages)

Did you know there are billions of years unaccounted for between verses 1 and 2 in the book of Genesis?

Genesis 1:1 - In the beginning God created the heaven and the earth.

1:2 - And the earth <u>was</u> without form, and void; and darkness was upon the face of the deep. And the Spirit of God moved upon the face of the waters.

I believe the key to understanding verse 2 lies in determining God's mood. We need to look beyond the words in order to understand the second verse—to read with the heart instead of the mind, which may actually better reveal the meaning behind God's words.

Try reading the words now, and see if you don't agree that God sounds somber, melancholy and almost remorseful. Later we will inspect every word, using this assumption, to see what is revealed to us. But first, the one thing we can determine for certain is that God is *not* creating the earth in this verse. He already created the earth in verse 1. Otherwise, he would *not* have used the past tense words "earth *was*". In verse 2, it's clear that the earth is already in existence.

Also, the word "And" is used, even though verse 2 is not dependent upon verse 1. Verses 1 and 2 are both independent statements. We know from the last chapter that astronomers have proven that the age of the earth is about four and a half billion years, and the work God did in six days, wasn't done four and a half billion years ago. The word "And" at the beginning of verse two confuses more Bible readers than any other word in the creation story. So, since each verse stands alone, when you read this verse, begin with the word "the".

Now let's study each word in verse 2 to see what may be revealed to us.

Without form is from the Hebrew word *tobu,* which means desolation or waste.

Darkness can have the connotation of evil, gloom, despair or desolate.

Void means empty, no longer filled.

Move upon can mean to brood over or to be moody.

Face of the deep suggests that the earth was flooded with water.

So it sounds like this:

The earth was in waste and no longer filled (with life?). Desolation was on the face of the deep. And the Spirit of God brooded over the flood waters.

It certainly sounds like God is revisiting the earth and assessing the damage done by a flood. Possibly, the memory of his necessary actions saddens him. Certainly God would never create an earth that was already desolate and in a chaotic condition. It is clear that the six days of "creation" is actually God doing the things necessary to make the earth habitable once again.

This is a God returning the earth to its original form and function. He's doing what has to be done before He can create man and the creatures of the earth again. I say again, because there is overwhelming evidence which shows that the earth was populated before the flood of Genesis 1:2. (To be covered later)

So, what we have always been taught to believe as God creating the earth in six days, is actually God *restoring* the earth in six days.

1:2 - And the earth <u>was</u> without form, and void; and darkness was upon the face of the deep. And the Spirit of God moved upon the face of the waters.

1:3 – And God said let there be light: and there was light.

The first work God did, in order to restore earth to its original condition, was to restore light to the earth. Here, he is not creating light, because the sun was already in place with the original creation in Genesis 1:1. Notice he didn't say, "I command that there be light to shine upon the earth". He is giving light permission to reach the earth. So he gave the command to *let there be* light. In earth's ruined state, apparently he had previously done something to block the sun's light from the earth.

God accomplishes his work by the spoken word, as in verse 3. When God initially gave the command that ignited the creation of heaven and earth, it took ten billion years before our solar system became a reality; and maybe another billion or so years for earth to become ideal to support life. When God commands something to occur, even his command is subject to His own natural physical laws of the universe. So, in the six days of restoring the earth to a habitable condition, each command may have taken hundreds, or even thousands of years to materialize.

1:4 - And God saw the light, that it was good: and God divided the light from the darkness.

1:5 - And God called the light Day, and the darkness he called Night. And the evening and the morning were the first day.

On the second day, God removed all excess water from the earth in the form of clouds (firmament) and placed them in the sky (heaven).

1:6 - And God said, Let there be a firmament in the midst of the waters, and let it divide the waters from the waters.

1:7 - And God made the firmament, and divided the waters which were under the firmament from the waters which were above the firmament: and it was so.

1:8 - And God called the firmament Heaven. And the evening and the morning were the second day.

On the third day, God restored the divisions between the dry land and the seas. He also brought forth vegetation.

1:9 - And God said, Let the waters under the heaven be gathered together unto one place, and let the dry land appear: and it was so.

1:10 - And God called the dry land Earth; and the gathering together of the waters called he Seas: and God saw that it was good.

1:11 - And God said, Let the earth bring forth grass, the herb yielding seed, and the fruit tree yielding fruit after his kind, whose seed is in itself, upon the earth: and it was so.

1:12 - And the earth brought forth grass, and herb yielding seed after his kind, and the tree yielding fruit, whose seed was in itself, after his kind: and God saw that it was good.

1:13 - And the evening and the morning were the third day.

On the fourth day, God restored regulation to the earth in relation to the universe, so that it would again have seasons. He did not do this on the first day, when he restored sunlight to the earth.

1:14 - And God said, Let there be lights in the firmament of the heaven to divide the day from the night; and let them be for signs, and for seasons, and for days, and years:

1:15 - And let them be for lights in the firmament of the heaven to give light upon the earth: and it was so.

1:16 - And God made two great lights; the greater light to rule the day, and the lesser light to rule the night: he made the stars also.

1:17 - And God set them in the firmament of the heaven to give light upon the earth,

1:18 - And to rule over the day and over the night, and to divide the light from the darkness: and God saw that it was good.
1:19 - And the evening and the morning were the fourth day.

On the fifth day, God gave his command to restore the creatures of both the sea and the air. He could only do this after He had prepared a way for them to survive, which He did in days one through four.

1:20 - And God said, Let the waters bring forth abundantly the moving creature that hath life, and fowl that may fly above the earth in the open firmament of heaven.

1:21 - And God created great whales, and every living creature that moveth, which the waters brought forth abundantly, after their kind, and every winged fowl after his kind: and God saw that it was good.

1:22 - And God blessed them, saying, Be fruitful, and multiply, and fill the waters in the seas, and let fowl multiply in the earth.

1:23 - And the evening and the morning were the fifth day.

On the sixth day, God finally restored the animal kingdom, and then re-created man after his own image. Pay special attention to verse 1:28, where God tells man to *"Be fruitful and multiply, and _replenish_ the earth...."*

In Webster, we read: *"re-plen-ish (ri-plen'-ish) v.t. to fill up again; to restock.*

Here, God's literal statement clears up any doubt that this is a "do-over" for mankind. There is no way to refill something, unless it existed before.

1:24 - And God said, Let the earth bring forth the living creature after his kind, cattle, and creeping thing, and beast of the earth after his kind: and it was so.
1:25 - And God made the beast of the earth after his kind, and cattle after their kind, and every thing that creepeth upon the earth after his kind: and God saw that it was good.

1:26 - And God said, Let us make man in our image, after our likeness: and let them have dominion over the fish of the sea, and over the fowl of the air, and over the cattle, and over all the earth, and over every creeping thing that creepeth upon the earth.

1:27 - So God created man in his own image, in the image of God created he him; male and female created he them.

1:28 - And God blessed them, and God said unto them, Be fruitful, and multiply, and replenish the earth, and subdue it: and have dominion over the fish of the sea, and over the fowl of the air, and over every living thing that moveth upon the earth.

1:29 - And God said, Behold, I have given you every herb bearing seed, which is upon the face of all the earth, and every tree, in the which is the fruit of a tree yielding seed; to you it shall be for meat.

1:30 - And to every beast of the earth, and to every fowl of the air, and to every thing that creepeth upon the earth, wherein there is life, I have given every green herb for meat: and it was so.

1:31 - And God saw every thing that he had made, and, behold, it was very good. And the evening and the morning were the sixth day.

2:1 - Thus the heavens and the earth were finished, and all the host of them.

2:2 - And on the seventh day God ended his work which he had made; and he rested on the seventh day from all his work which he had made.

2:3 - And God blessed the seventh day, and sanctified it: because that in it he had rested from all his work which God created and made.

In reading the Bible with this new understanding of Genesis 1:1 through 2:3, we totally dissolve one of the age-old standing differences between science and religion regarding the age of the earth. Scientific evidence of a 4.5 billion year old earth actually supports the possibility of there being a divine creator. We read in Genesis 1:1 In the beginning *(which we now know means billions of years ago)* God created the heaven and the earth.

In God's infinite wisdom, He has worded Genesis 1:1 through 2:3 in such a way as to enable Moses (early man) to have a simple view of his "center of the universe" world, while giving 21st-century man the clues needed to reveal the truths about creation and re-creation. This opens an entirely new perspective on the existence of a "Pre-Adamite" world, which would certainly warrant further investigation into the evidence uncovered by modern science's archeological digs.

So, let's go on an expedition into the murky, dateless past, to try and answer some of life's most thought-provoking and intriguing questions.

Did you know . . .

. . . that nowhere in the Bible does it claim that Adam was the first man on earth?

. . . that the existence of prehistoric man is not contradicted in the Bible?

Chapter 5

Digging Up the Past

In the Biblical account, there can be no doubt that there was an advanced social order, which existed on the earth prior to the creation of Adam, and that much of their existence was violently and abruptly disrupted through the fulfillment of a worldwide, flood as a result of God's wrath.

The flood of Genesis 1:2 is not to be confused with Noah's flood, which didn't occur until 2348 B.C. However, there is definitely a connection between the two floods. . . and that connection is Lucifer, also commonly known as Satan, the Devil or Beelzebub. I will cover this in detail later.

Lucifer is never spoken of in the present tense during the period covered by the Old Testament, so we know that he had to exist sometime in the dateless past, before the creation of Adam. This is an additional reason to believe that there was life on earth prior to Adam. From scripture, we learn that Lucifer was God's trusted angel, whom He had left in charge of ruling the earth. There is insufficient Biblical information to determine how long Lucifer ruled the earth. However, we do know that eventually he fell from God's grace due to his extreme self-pride and wickedness. (See Ezekiel 28:15)

In retaliation, Lucifer led a coup to overturn God's authority by invading Heaven, using one-third of God's own angels. This, of course, was a spiritual war, which I'm certain is extremely more severe than any physical war that could be waged. According to scripture, Lucifer and his rebel angels were defeated and banished to the earth, thus the term "fallen angels." More than likely, Lucifer was an invisible ruler and not seen by his earthly subjects, just as we cannot see God. So, just because Lucifer ruled the earth, does not mean that civilizations of the earth would have been aware of his presence.

In order to give credence to the existence of these pre-Adamite inhabitants of earth, it would behoove us to first establish the dating of God's creation of Adam. Since God hasn't chosen to disclose this information through His inspired word, one can only estimate by backtracking, and using genealogy from a suspected descendant, which in this case would need to be Jesus of Nazareth.

Doing a genealogical study always reminds me of my great Aunt Etta, my grandfather's sister, who made it her life's mission to trace the roots of our family name. As a child in the 1940s, I can still vividly recall her proudly unveiling the decades-long work one evening in my grandparents' living room. She had successfully traced our family's roots back to a soldier who had fought in the battle of Hastings in 1066 A.D. I was amazed by her accomplishment of uncovering nearly 900 years of history, using only a pen and paper, and who knows how many postage stamps.

Today, through the Internet, with access to the Mormons' database, as well as nearly every record-keeping institution around the world, the same feat could probably be achieved in a matter of weeks. But tracing family roots in Biblical times, when writing instruments were crude at best and data research involved the use of a camel, doing a lengthy genealogy would have been an extremely difficult, if not impossible, undertaking.

Some seventy years after Christ's crucifixion, a Bible writer who identifies himself simply as Luke, wrote the third of four Gospels in the New Testament. "The Gospel According to St. Luke" is an account of Christ's life. Luke has also been given credit for writing the book of Acts, because the writing styles in both books are so similar. Just who Luke was, has had Bible scholars and critics alike speculating for decades. This much is fairly well agreed upon:

Luke was probably a highly educated man, well traveled, and in keeping with the times, wrote in the Greek language.

It is widely believed that Luke collected much of his unique material during the imprisonment of Paul in Caesarea, when Luke attended to him. Paul described Luke as "the beloved physician," so he could have had medical training. This particular Gospel is directed toward a Gentile audience, making the point that Christianity is not a Jewish sect. So Luke himself was probably a Gentile.

He has addressed the Gospel to "Most Excellent Theophilus," which could either be a reference to the Roman-imposed High Priest of Israel between 37 and 41 A.D., or it may simply be the Greek generic term for Christian "loved by God."

But regardless of who Luke was, one of the most amazing disclosures that he contributed was Christ's complete genealogy, starting with Joseph and going back 77 generations through Noah to end with Adam, God's created son. It was finally James Ussher, Ireland's Archbishop of Armagh (1581–1656), who calculated the date of Adam's creation to be the night preceding Sunday, October 23, 4004 B.C.

Being a prolific scholar and knowing Christ's generations from the Book of Luke, Archbishop Ussher was able to piece together enough clues found in the Old Testament to establish the exact date of creation. Biblical scholars from more recent times have re-calculated the event and found Ussher's date to be correct.

By the end of Lucifer's pre-Adamite world, we found an earth which lay in ruin and was desolate. While the length of Lucifer's reign cannot be determined, from scripture we can be certain that the earth would have been covered with water for a period of time before God restored it to a habitable condition, as seen in Genesis 1:2 through 2:3. If this is true, there should be evidence of a flood sometime prior to the creation of Adam in 4004 B.C. Since this predates the recorded history of all the world's civilizations, one can only depend on the findings of present-day science to determine if there is evidence of a flood or deluge during the time in question.

In researching this part of my book, I found that there were actually three separate climate events that could have contributed to the flood of Genesis 1:2.

Let me re-emphasize that I am not addressing the flood of Noah's time, which occurred in 2348 B.C. I am addressing the earth's flooded condition as described in Genesis 1:2, when "the spirit of God moved upon the face of the waters." I feel that I must make this distinction again due to the automatic brain response at the utterance of the word *flood*. Whether legend or reality, it is the story everyone is taught from a very young age.

As I typed the words "6000-year-old flood" into Google's search engine, all kinds of references to Noah's flood kept coming up, but nothing for Adam's flood of Genesis 1:2. There were even some contributors claiming floods from thousands of years earlier than Noah's flood, totally ignoring the chronology of Noah, just to make the find fit the story. It's like the old saying, "If the only tool you own is a hammer, every problem begins to look like a nail." Obviously, very few people are even aware that the earth was flooded just prior to the creation of Adam.

The first climate event that may have contributed to the flood of Genesis 1:2 was *The Neolithic Sub Pluvial*. As with most scientific terms, this one could use the skills of an interpreter, so allow me: Neolithic means late Stone Age. The prefix "sub" refers to the ground, and pluvial simply means rain. So, Neolithic sub pluvial is the scientific way of saying that during the late Stone Age, the ground was saturated with rainwater.

This is the geological condition that existed in the Cradle of Civilization of northern Africa, where the world's early population was concentrated. This condition prevailed from about 7500 B.C. to about 4000 B.C. Although this was both preceded and followed by much dryer periods, for nearly 4000 years, the Sahara desert was a rain forest.

This was followed by the second event, the melting and draining of *Lake Agassiz*. Lake Agassiz was a gigantic glacial lake located in the center of North America. Ten to twelve thousand years ago, it included areas of Manitoba, Ontario, Minnesota, North Dakota and Saskatchewan. It may have covered as much as 327,000 square miles. While it partially drained and refilled with melting ice several times during the last Ice Age, along with climate changes, the final shift in drainage in about 6200 B.C. added from four to ten feet to global sea levels. Geologists believe that this last drainage could have occurred in a period as short as a couple of years.

The final climate event was called *The Older Peron Transgression.* This was simply runoff from glaciers, caused by a period of unusually warm climate conditions that lasted from 4900 B.C. to 4100 B.C. This was a global event that added eight to thirteen feet to sea levels for a period of eight hundred years. This anomaly was named for Cape Peron, along the west coast of Australia, where terracing of the coastline caused by water erosion can still be seen today.

We have all seen what a devastating force water can be when unleashed against an unsuspecting population. Only recently, in a freak storm in Australia, six inches of rain fell in just thirty minutes, creating a flash flood that swept away dozens of people and destroyed hundreds of homes. This storm sent a ten-foot high wall of water crashing through the streets of Toowoomba, carrying away anything in its path toward the ocean.

This was the result of a localized storm, with modern-day communication and rescue technology in place. Can you imagine the combined effects of these two global events in tandem, with no technology of any kind? This should have destroyed all of humanity. And then there is the resulting pestilence from rotting corpses. With the low survival skills of primitive peoples, the residual effects of disease would have run rampant, finishing off would-be survivors in great numbers.

Was this the end of humanity? Despite the devastation God placed upon the earth, I am confident that a certain number of people from the older creation did survive. To support this, digs have revealed human remains throughout this period of time.

When I began doing the research on this book, I didn't really know what I would find. I knew what I wanted to find. I wanted to find proof that religion and science were both right, but I knew that was an unrealistic expectation. Having grown up in a Christian home, I had developed a deep respect and love for God, but I also had an inquisitive fascination for science. So much of what I learned about science in school was based on logic, and that appealed to me. I just always accepted everything in my science textbooks as being the final word of authority. So when the big fork-in-the-road moment came for me, I felt immense frustration.

I'll always vividly remember when our junior high school class was first introduced to the workings of the human heart. The full color pictures in my textbook looked so real. I suppose I knew they were only an artist's rendering, but to a thirteen-year-old, it looked like someone had actually cut this dead person's heart out and sliced it open, so you could see the inside of every part of it, even the arteries.

I was so impressed with that section in the book that I memorized the names and functions of every part of the heart. In my spare time, I would even draw and label them. I prided myself in being able to describe how the blood was pumped from one side of the heart, through the arteries and veins of the body, and then would go through the lungs to be re-oxygenated before returning to the opposite side of the heart, just to start the cycle all over again.

Our teacher, Mr. Whitlock, made learning fun. Everyone tried to arrange their classes so they could have him for science. He had a flamboyant personality and was probably in his mid-thirties, short in stature and plump, but he was always smartly attired. A forest green dress shirt, rust-colored tie and white lab coat were typical.

The lab coat was funny because there was no science lab in the school. So he would occasionally conduct a science experiment in the classroom and involve the students. Of course, these experiments usually ended in some unexpected, grand finale. Oddly, the principal always seemed to be away from school when he conducted his random experiments.

When we had a test at the end of that section of the book, I made the only A-plus in the entire class. This was anything but typical for me. When Mr. Whitlock handed the graded papers back in class, he commented on my achievement. Everyone in class stared in disbelief. They knew me. And then Mr. Whitlock asked, "Well, class, before we leave this section in the book, does anyone have any questions?"

Now, I wasn't one to draw attention to myself. My theory was, if you kept your head down and didn't ask any questions, you couldn't get laughed at for asking something stupid. But this day, I wasn't going to take my own advice. I had a serious question, and I was dying to know the answer from an adult. I nervously cleared my throat. Every eye in the class was on me again. The teacher said, "Yes, Bob, do you have a question?"

It was too late to back out now. I felt everyone staring at me, waiting, probably thinking, "What's the question, nerd? I thought you knew it all." Finally, I found my voice.

"Mr. Whitlock, I know how the heart works, but *how* does it know to beat?"

"It knows to beat because the brain regulates that function. But we've already covered that."

"I know that, but I mean....why are we alive in the first place?"

I thought I noticed Mr. Whitlock's face turning red. Slowly, he peered over the top of the horn-rimmed glasses perched on the tip of his nose. He sort of glared at me as he raised an eyebrow—not both, just one. The silence that followed seemed to go on forever. Had he heard me? Did he want me to repeat my question? Finally, with a sly grin, he answered.

"Mr. Follett, I don't know if you've taken noticed of this, but all semester you've been sitting in a science class, not a theology class. You're going to have to ask questions like that of someone else." With that, he pushed his glasses back into place with one finger. The whole class chuckled.

"Does anyone else have a question?" No one did.

I could feel my face suddenly heat up with embarrassment. Now, I wasn't even sure what theology was, but my one question that went unanswered taught me more than any answer he could have given. To me, this was monumental. Science seemed to have all the answers to the "what" questions. It's the "why" questions they couldn't answer.

When I discovered that geologists had found evidence of great flooding in the earth's strata, covering a 2000-year period ending just before the fourth millennium B.C., I knew the "what."

But when I compared that with the time of Adam and Eve's creation, I knew the "why." I could only conclude that scientific evidence does, indeed, support the existence of a divine Creator. I knew that this was a very significant concord between science and religion.

I believe the chart below will allow you to visually comprehend the devastation of this 2000-year-long flood of the earth.

Did you know . . .

. . . that giants inhabited Canaan *(the promised land)*, and that they were offspring of fallen angels who had fornicated with earthly women? You can read about it in Chapter 6.

Chart #3
2000 year flood of Genesis 1:2

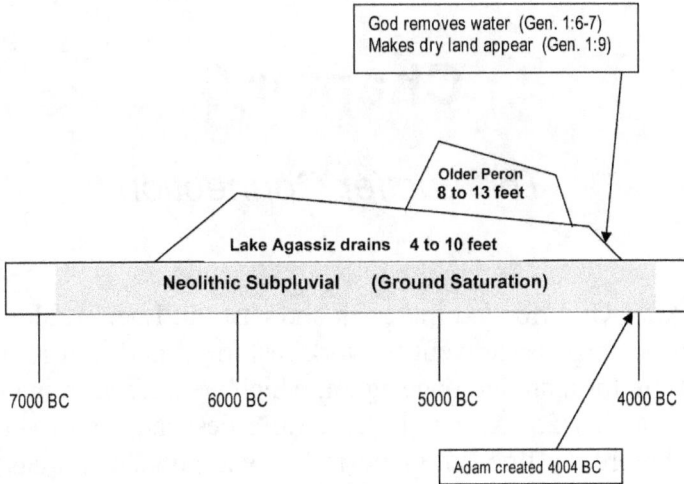

God removes water (Gen. 1:6-7)
Makes dry land appear (Gen. 1:9)

Older Peron
8 to 13 feet

Lake Agassiz drains   4 to 10 feet

Neolithic Subpluvial   (Ground Saturation)

7000 BC   6000 BC   5000 BC   4000 BC

Adam created 4004 BC

Chapter 6

The Lucifer Connection

After God flooded the earth and cast out Lucifer, along with his fallen angels, He went to work restoring the Earth to make it habitable for mankind once again, which we have seen in Genesis 1:2 through 1:25. Although there were descendants of survivors from before the flood of Genesis 1:2 who populated areas of the Earth, God prepared a place in Eden for a new kind of man—a super human—formed after the appearance of God himself and physically capable of living for nearly a thousand years. Here in Eden, God planted a garden for man to tend, and gave him a wife, whom Adam later named Eve.

The Garden of Eden

It's only natural that mankind would want to know the exact location of his origin. And while science has convincingly named northern Africa as the area where mankind originated, known as *the Cradle of Civilization,* Biblical scholars have been trying for hundreds of years to unravel the clues given in the book of Genesis regarding the location of the Garden of Eden, where this new super race of humanity had its beginnings.

Unfortunately, after hundreds of years of deciphering, the Near-Middle-East is about as close as they have been able to pinpoint the location of this mysterious garden. And there's good reason for this. The clues given in Genesis are not clues at all— they are undecipherable phrases intended to conceal the location of Eden, and they have served their purpose well. We should all face the fact that there are certain things we are just not meant to know, and this is one of them. That is not to say that there haven't been some good educated guesses, but without proof, that is all they are.

Printed below are the Biblical clues given in Genesis 2:10-14, followed by my reasons why they are totally unworkable in locating the Garden of Eden.

¹⁰And a river went out of Eden to water the garden; and from thence it was parted, and became into four heads.

There is no river in the Near-Middle-East that divides into four rivers.

¹¹The name of the first is Pison: that is it which compasseth the whole land of Havilah, where there is gold;
¹²And the gold of that land is good: there is bdellium and the onyx stone.

There is no river named Pison, nor is there a land that has ever been known as Havilah. Neither name appears anywhere else in the Bible.

¹³And the name of the second river is Gihon: the same is it that compasseth the whole land of Ethiopia.

There has never been a river named Gihon, and there is no river that encompasses all of Ethiopia. Ethiopia is southwest of the Persian Gulf, opposite the side anyone believes Eden to be on.

[14]And the name of the third river is Hiddekel: that is it, which goeth toward the east of Assyria. And the fourth river is Euphrates.

There is no Hiddekel River near Assyria, or anywhere else known of throughout recorded history. The Euphrates is the only one of the four rivers named that exists today, but still, it is not a division of three other rivers.

Some Biblical scholars have suggested that perhaps the names have changed from what they were in Adam's day, but that claim is without foundation, because God gave Moses the text of Genesis in Moses' time. There is nothing to support the theory that the names of rivers and other geographical landmarks would have changed since then.

I believe it's clear that Eden's location is to remain a mystery. And does it really matter that we don't know where it was located? After all, isn't God telling us all that we really need to know?

Everything was perfect in the Garden of Eden for a new beginning. But it didn't last long. Lucifer, in retaliation to God, appeared as a serpent to Eve *(See symbolic interpretation at Chapter #7)* and convinced her to eat the fruit of the tree of knowledge of good and evil. And of course, the rest is history. When God created Adam and Eve, He created them as adults. After they were cast out of the garden, God placed Cherubim (heavenly creatures) east of Eden to keep others from eating of the second and more important tree, the Tree of Life. The fruit of this tree would have given immortality.

Adam and Eve began to have children. And while Cain, Abel and Seth are the only offspring mentioned by name in the Bible, they also had other sons and daughters (see Gen. 5:4). In order to continue this new superior race, it was necessary for the sons of Adam and Eve to marry their sisters, which they did.

This is easily confirmed in Genesis 3:20, in which we read:

And Adam called his wife's name Eve; because she was the mother of all living.

If Eve was the mother of all living, then it follows that brothers had to have married sisters. But God designed this new species of mankind so perfectly that what we consider incest today did not corrupt man's genetic makeup.

Please take a moment to study the following chart before continuing. It will help you understand the chronology of Lucifer's attempts to block Christ's arrival.

Chart #4
Royal Ancestry From
Lucifer's Reign to Jesus' Birth

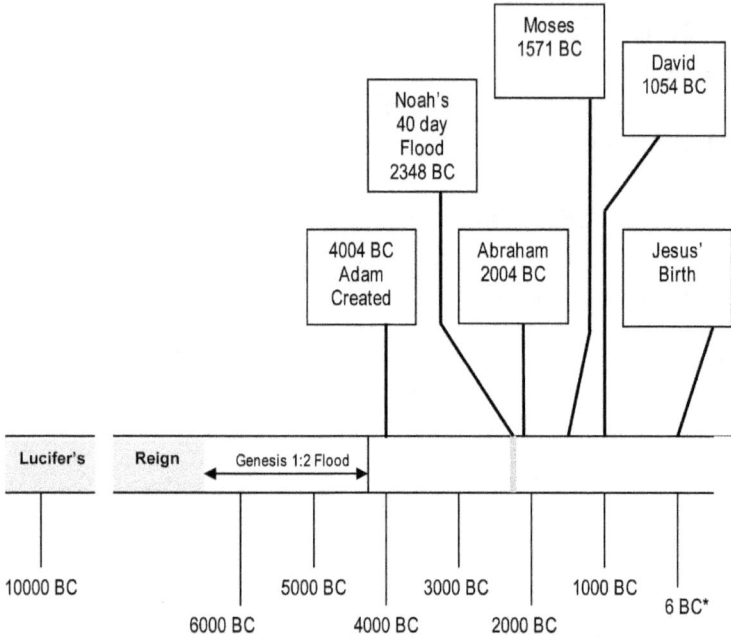

While there were seventy-seven generations from Adam's creation to Jesus, this chart shows only the major patriarchal figures, placed so you can readily understand the chronology of Jesus' ancestry in relation to the events referred to throughout this chapter.

* It has been established with some certainty that
Christ was actually born April 17, 6 BC.

Lucifer didn't stop with the trickery he performed on Eve. He tried to keep Jesus from arriving as man's savior multiple times, and in various ways, both before and after Noah's flood. I have previously mentioned one of these occasions early in Chapter 3 when Lucifer, working through the Pharaoh, slew all the Hebrew babies in captivity in an attempt to kill Moses, who was in Christ's lineage. Later, after the birth of Christ, he made another similar attempt through King Herod by having all the Hebrew babies under the age of two destroyed, but Joseph and Mary were warned by the wise men and escaped to Egypt.

These are just two examples of the many times Lucifer tried to stop Christ by destroying *the seed of the woman.* Another method he used was to corrupt the pure bloodline through which Jesus was to be born. In order for the prophecy of Christ arriving as man's savior to be fulfilled, there needed to be continuous uncontaminated genetics, traceable back to Adam. The Biblical phrase for this is "one who is perfect in his generations."

The Giants

Lucifer tried to keep Christ from arriving as the Messiah by having his fallen angels fornicate with earthly women. These unions resulted in the birth of evil giants who grew to be 10 to 30 feet tall. Goliath is probably the most notable among them, but he was only one of many to plague mankind. This not only occurred before Noah's flood, but would also continue after the flood. In Genesis 6:4, we read:

There were giants in the earth in those days; (before Noah's flood) and also after that, (after the flood) when the sons of God (fallen angels) came in unto the daughters of men, (earthly women) and they bare children to them, the same became mighty men, which were of old, men of renown.

After several generations, this contamination would aggressively promote the spread of sin and wickedness throughout the entire population of the Earth, and thus thwart God's plan for the coming of Christ as man's savior.

Now, I have a serious question for you. What just registered on your believability scale when you read the word giants? Like me, didn't you think giants were found only in fables—the imaginings of delusional storytellers, fictional subjects of awe, whispered by children sitting around a campfire? How long has it been since you seriously thought about giants? I believe the earliest memory I have of them is of someone reading the story "Jack and the Beanstalk" to me.

Of course, almost immediately we're taught that giants aren't real. After all, we wouldn't want any of them showing up under our bed or in our closet after everyone else is asleep, would we? So, we go through life continuing to believe that giants only exist on the pages of a child's storybook. The fact is, however, that giants really did exist. And the proof of their existence can still be found today.

Now granted, those findings are elusive, but only because the fascination with giants exists today to such a degree that it causes many to perpetrate elaborate hoaxes in order to convince the non-believers. As a result, one is unable to determine the difference between fact and fiction.

In my research on giants, I have found many photographic images, which after extensive research and careful examination have normally turned out to be someone's Photoshop creation. These images generally look so real that they defy detection. I have also found testimonials regarding scientific entities claiming to have located graves of giants measuring 6 meters long (19.7 feet), or fossilized footprints of gigantic proportions. But when you try to authenticate the source of the find, they all just fall short of disclosing anything that would definitively establish the claim as positive, complete, factual truth.

So, I have decided to throw out all photographic images, disregard all Internet claims, and work only with what I consider to be evidence originating from unquestionable, authentic sources. Believe me, this eliminates at least 95% of what is out there to find.

My first and most reliable source of information about giants would have to be from the Holy Bible. Let me explain why. Every scientific field dealing with historic findings always give

100% credibility to any form of recorded history they find. It doesn't matter whether they locate recorded history carved on the walls of caves, as cuneiform writing on Egyptian tombs, or in etchings on ancient clay pottery—they accept it as reliable scientific evidence and fully regard the content of those writings as factual.

However, when the *Dead Sea Scrolls* were found, which confirmed sections of the modern-day Bible, the response of the scientific community was nonexistent. Even though the Bible has been proven accurate time after time, they have turned a blind eye to any Biblical find that contradicts established scientific beliefs.

When I Googled *Scientific evidence of giants,* I found nothing concrete—only dead-end claims that couldn't be substantiated, which didn't really surprise me. According to the Bible, Lucifer's giants only existed for a few hundred years, during a period before and after Noah's flood. Compare that to the two million years that humans, of one type or another, have been found to exist on the Earth. The chances of finding giant remains would be far less than an individual winning back-to-back multimillion-dollar lotteries. But I must assert to you that the lack of findings never proves that something didn't exist.

In addition to the Bible, another source of information, which I also consider to be irrefutable, is that information given to us by Flavius Josephus, the famous Roman historian. Flavius was a Jewish leader who was captured by the Romans and became a Roman citizen in 67 A.D. We would know far less about the workings of the Roman Empire, had it not been for the prolific writings of Flavius Josephus.

My third source of information is the Dictionary of Phrase and Fable by Brewer. However, I have chosen not to use information from this source to originate a claim, but rather, only to confirm or clarify a claim from one of my first two sources. In addition to these, any other source I use will be specifically named.

To be totally honest, I had not even thought about giants since early childhood. At some point I must have simply filed the subject away in my mind as another of those mythical creatures in literature used to create a sense of awe and wonder in the minds of readers. I was more than willing to leave it neatly categorized as fiction and stored away in the recesses of my mind, content to consider it resolved.

But then, something very curious recently came to light that compelled me to reconsider what I already thought I knew. While I had closed the coffin lid on the subject of giants and buried it years ago, something was refusing to stay dead. I knew the time had come for me to exhume my old beliefs about giants and examine them once again.

This whole thing began about three years ago, when I received a mailing from one of those mail-order companies that never seem to run out of once-in-a-lifetime collectible offerings that you simply can't live without. Most of them go directly into the trash without even being opened, but this particular offer rekindled an old interest of mine, which for one reason or another I had always dreamed about but never pursued: that of collecting coins, and more precisely, ancient coins.

I guess I'm somewhat of a nut when it comes to anything old. In visiting a museum, I can stand before a single display and become so mesmerized by its historical significance that when I finally awaken from my awe-induced trance, the rest of the group is three exhibits away.

The mailing I received showed three ancient Roman coins, each encased in its own clear, acetate bubble, attractively displayed on a storyboard about the Roman emperor who authorized that coin's minting. For only twenty dollars per month, plus shipping, I would receive an ancient Roman coin every other month, making each coin's cost about forty dollars. The genuine leather display binder would be sent free of charge upon receipt of the third payment. I was impressed that you could own such an awesome piece of history for only forty dollars. I checked the offer against the same items offered on eBay and found that, as a hobby, coin collecting could be as affordable as it was intriguing. I was hooked.

The first thing I did was to order the book *Handbook of Roman Imperial Coins,* by David Van Meter, so that I could have a better understanding of coin values. This book is laid out in sections covering the various time periods of the Roman Empire. At the beginning of each section, there is a biographical synopsis outlining every significant occurrence that is known about the various emperors who ruled the Roman Empire during that time period.

As I was poring over this information one evening, the name of one of these emperors jumped out at me: Maximinus. His official Roman name was Caius Julius Verus Maximinus, and sometimes the name Thrax is attached, but he became known in history as simply Maximinus. Allow me to quote from this extremely fascinating book.

"Maximinus was a herdsman before joining the army of Septimius Severus. Due to his enormous physical stature, as well as his military skills, he rose through the ranks to legionary command..."

As I read this, I wondered how enormous one's physical stature would need to be in order to be remembered and recounted throughout nearly eighteen hundred years of recorded history.

I turned to another book in my modest library, which I had obtained as a gift from my stepdaughter when she learned of my interest in Roman coins: *The History and Conquests of Ancient Rome* by Nigel Rodgers, Consultant: Dr. Hazel Dodge, FSA. On pages 78 to 79, I read the following:

JULIA MAMMAEA – When Alexander Severus succeeded his assassinated cousin, the depraved Elagabalus, to become emperor in AD222, he was only 14 years old and under the sway of his mother, the Syrian princess Julia Mammaea. He never managed to escape her maternal domination, but at first Julia ruled very effectively. She reversed all Elegabalus' scandalous policies; chose 16 distinguished senators as advisers and relied heavily on the famous lawyer Ulpian, also from Syria, whom she made commander of the Praetorians. However, Ulpian proved unable to control the Praetorians, and was finally murdered by them in AD228.

*Meanwhile, Julia had become madly jealous of her son's wife, Barbia Orbiana, whom Alexander had married in AD225, and whose father he had made Caesar or co-ruler. Julia had her daughter-in-law thrown out of the palace and her father executed. Being still dominated by his mother, Alexander accepted this, but Julia could not control foreign attacks. After an inconclusive expedition to repel a Persian invasion in AD232, mother and son went north to deal with a German attack. Alexander so alienated the Rhine legions by his military feebleness and his meanness about their pay that they chose the **giant Maximinus** as emperor in AD235. Troops sent to kill Alexander found him clinging to his mother in a tent. Mother and son were butchered together, so ending the Severan dynasty.*

When reading the account of the giant Maximinus, I turned to the *Dictionary of Phrase and Fable* by Brewer, and also found the following on page 397:

Maximinus I was 8 ft. 6 in. in height. Roman emperor from about 235 to 238.

Finding corroborating evidence in three separate, reliable sources left no doubt as to the existence of giants—or at least, this giant.

I have a good friend who is the size of many of the larger NBA players. He stands a full six foot, eleven inches tall, and although he's never shown an interest in basketball, saying no to his many would-be coaches has consumed a lot of his time. Whenever I see Larry and shake his hand, for just an instant I'm transported back in time to my five-year-old self, holding my dad's hand while crossing the street. I'm always amazed by how huge Larry's hands seem, even though they're in proportion to the rest of his body.

Larry is fourteen inches taller than I am—but imagine someone nineteen inches taller than Larry. If my friend could somehow shake the hand of Maximinus, would he experience this same I'm-feeling-like-a-child phenomenon? I'm fairly certain that he would.

And then my research brought me to an even larger giant than Maximinus. His name was Eleazar, and he was a Jew living during the time of Jesus. Although this giant is never mentioned in the New Testament, his existence is well documented in the writings of the Roman historian Flavius Josephus. This was during the reign of the Roman emperor Tiberius, who had become quite paranoid in the last decade of his rule. In order to support his reclusive manner, had moved his court to the fortress island of Capri in 27 AD. Here he ruled very harshly by proxy through his many functionaries. Principal among them was Vitellius, whom Tiberius had made president of the Roman province of Syria, headquartered in Antioch.

According to Josephus, the giant Eleazar came as one of several gifts exchanged between Tiberius and Artabanus, king of Parthia, during a truce that Vitellius had arranged. This was a result of wars fought between Rome and Parthia over Armenia, which had been both lost and then regained by Rome. But after monetary bribes given to the kings of neighboring countries to kill Artabanus had failed, and defeating him seemed an elusive option, Tiberius had Vitellius arrange a truce. Josephus reports Eleazar's height at seven cubits, which would be nearly eleven feet tall.

I found the Dictionary of Phrase and Fable by Brewer to give the following information on page 397, under "Giants of Later Tradition."

"Eleazer was 7 cubits (nearly 11 ft.). Vitelius sent this giant to Rome; he is mentioned by Josephus. Goliath was 6 cubits and a span."

So, let me put forth this reasonable assumption. Since practically everything we know about the Roman civilization originated from the writings of Josephus, and we have accepted these writings as factual regarding all other aspects of his life and times, we cannot now begin to pick and choose which of his statements about giants we find believable based on our own narrow understanding.

If we were to reduce everything we consider to be true down to only those things that fit neatly within the confines of our own preconceived beliefs, then on this same basis, we may as well throw out all of mankind's recorded history.

Continuing with this line of reasoning, since nothing written in the Holy Bible as being literally stated has ever been proven wrong, we must accept the verses written about giants to be factual as well. As with Josephus' writings, either the Bible is entirely true, or it is entirely false. There is no in-between. When we begin to pick and choose which verses we believe, the entire Bible comes into question as a reliable historical document.

I mentioned before that Lucifer's giants were present on Earth both before and after Noah's flood as an attempt to thwart the coming of Christ. Of course, the entire purpose of God's flooding of the Earth was to rid it of these evil beings. Only Noah and his family, whose bloodlines had not been corrupted, were kept alive in the ark and lived after the flood. So the eruption of giants after the flood was a new outbreak, which Lucifer orchestrated against Noah's descendants. This second outbreak began around Moses' time, but would not be totally extinguished until David's reign as king.

Although giants are referred to many dozens of times in the Bible, the average reader would not necessarily be aware of these references, because they are actually called giants in only a few instances. While Nephil (meaning bully, tyrant, or giant) is the accepted translation for giant. they are often referred to by their tribal names, such as Rephaim or Anakim. These were the tribes of giants inhabiting either side of the Jordan River in the land of Canaan when Moses sent the ten spies ahead to investigate. On their return, they reported that they were mere grasshoppers by comparison. (*See Numbers 13:33*)

Another reason a reader would likely miss a reference to giants are in those instances when a giant is called either "dead," "death" or "deceased." It is clear in scripture that since these giants were not of pure human stock, they had no resurrection (see Isaiah 26: 14-19). And because they would not rise again, they were also referred to as the dead. In fact, Rephaim in Hebrew is translated as deceased or dead. One may confirm this fact by reading these passages: Proverbs 2:18; 9:18; 21:16 and Isaiah—14:9, where giants are referred to as "dead," and Isaiah 26:14, where they are referred to as "deceased."

You may be surprised to learn that the 23rd Psalm, which I'm sure you have recited many times, particularly if you have attended many funerals, is actually a reference to the land of giants. "Yea, though I walk through the shadow of the valley of death, I shall fear no evil." Giants were exactly the evil that David feared, and walking through their valley would have certainly been fearful. In this psalm, David was undoubtedly referring to the valley of the River Jordan, which runs through Canaan, the promised land.

You will notice above that the *Dictionary of Phrase and Fable* listed the infamous Goliath, whom David slew, as being six cubits and a span in height. By my understanding of these ancient units of measure, this would have made him nearly ten feet tall. But the giant ruler of Bashan, named Og, is mentioned in the Bible as having an iron bed nine cubits long and four cubits wide (*see Deuteronomy 3:11*). This would have made Og at least twelve feet tall—twice the height of modern man.

So, there you have it. God's reason for bringing on a catastrophic flood so severe that the story of it would be told and retold throughout time, passing from generation to generation to this very day, was to rid the world of Lucifer's population of evil giants. Although these evil beings would be back in the land of Canaan centuries later, attempting to prevent the Israelites from accepting God's gift of the promise land, God decided to destroy his creation a second time in order to keep Lucifer from infecting the seed of the woman, thus preventing the coming of Christ as man's savior.

In Genesis 6:5-7 we read:

⁵And God saw that the wickedness of man was great in the earth, and that every imagination of the thoughts of his heart was only evil continually.

⁶And it repented the LORD that he had made man on the earth, and it grieved him at his heart.

⁷And the LORD said, I will destroy man whom I have created from the face of the earth; both man, and beast, and the creeping thing, and the fowls of the air; for it repenteth me that I have made them.

Did you know . . .

. . . that descendants of those who survived the flood of Genesis 1:2 were symbolized by the serpent who convinced Eve to eat of the forbidden fruit? They would have been present in the area of the Garden during Adam and Eve's time there. This is explained in the next chapter.

Chapter 7

Noah's Ark

There are several Bible stories from the Old Testament that I think most people find very difficult to believe.

Many people question whether God actually parted the Red Sea for Moses during the Israelites' exodus from Egypt. Did the walls of Jericho really crumble and fall at the sound of Joshua's trumpets? These are only two of many stories that I believe could test the rationale of the more highly evolved, independent thinkers of today.

During my lifetime, I have heard more people voicing skepticism regarding the story of Noah's Ark than any other Biblical narrative. How could a vessel have been constructed, large enough to accommodate both male and female genders of the entire animal kingdom, not to mention having room enough to store the food required to feed such a troupe for an entire year? And even more inconceivable, how could Noah and his family have possibly gathered two of every known specimen of Earth's life forms into the Ark to make this yearlong voyage?

Certainly, no one living today could accomplish such a feat, even if they had the complete cooperation of every known animal conservation group, along with the help of both UPS and FedEx in order to expedite the handling and shipping aspects of such a venture. So how could Noah and his family, living over four thousand years ago, have possibly pulled off such an unbelievable achievement? Wouldn't the scientific community have fun refuting this story?

But hold on . . .Let us first dissect and carefully examine every aspect of this well-known Bible story to try and determine whether there could possibly be any truth in it at all. There are, after all, only two possibilities. This story is either fact or fiction. It can't be both. Right? So which is it?

First and foremost, we must remember that the book of Genesis, in which we find the story of Noah's Ark, is the only book in the entire Bible that was written in retrospect. That is to say, it was given to Moses after the fact. Everything in the book of Genesis had already happened before Moses was even born. So, God would have relayed this story to Moses in a way that would have been understood from Moses' viewpoint.

Allow me to reiterate what I have said before. God wasn't speaking to you or me—He was speaking to Moses. I'm certain that if God were telling today's man the story of His destruction, and then the restoring of all living things, He would tell it quite differently. But before we discuss the story of Noah's Ark, we need to consider the Old Testament text leading up to that point.

Old Testament teachings

Various Hebrews, who were inspired by the word of God, wrote the Old Testament in the Hebrew and Aramaic languages. These 39 books were written specifically for the Hebrew people. The Old Testament tells the history of the Hebrews, describing the struggles they endured and their ever-changing relationship with God.

In 1380 AD, when these original Hebrew manuscripts were translated into English, the Old Testament was changed from being scriptures about the Hebrew people, into being scriptures about the Christian ancestry to Jesus Christ. This unfortunate shift in emphasis has been responsible for more misunderstandings about the Bible than any other event in the history of religion. Today, after 630-plus years, this ancient Hebrew text is still being used to teach Christianity from a Christian viewpoint. As a result, entire nations of Christian-believing people have not only been grossly misinformed, but cannot even answer the simple question, "Who was Adam?"

If you ask any minister, pastor, reverend, priest or other church leader, including those who have doctorates in theology, from the lowest layperson to the Pope himself, you will be given the wrong answer. They will all tell you that Adam was the first man. The fact is; he most definitely was not.

Once I give you the real answer to this seemingly simple question, you will be amazed by how an erroneous belief, which has been allowed to exist unchallenged for centuries, has inadvertently thrown a cloak over the truth and changed the entire historic outcome, from fact to one of fiction. And the real wonder is, millions of people have believed it.

Knowing who Adam was, because so much in religion is based upon this knowledge, will give you a new, clearer view of the Old Testament. Suddenly, the inconsistencies will melt away, and you'll be able to make perfect sense of what was totally confusing before.

Unveiling the Truth

Let's begin with a statement of fact, which I know everyone can agree on. That fact being Jesus was a Jew—or more accurately, a Hebrew. There are 77 generations connecting Jesus to Adam in his ancestry. This can be verified in the Book of Luke. Going backwards from Jesus, and for the sake of brevity, considering only the patriarchal figures in his ancestry, let us begin with David. I think you will agree that since Jesus was

Hebrew, David would have been Hebrew as well. The next patriarch back was Moses. By the same reasoning, Moses would also have been Hebrew. And preceding Moses was Abraham, also Hebrew. Then there was Noah, who was Hebrew as well.

And finally, we come to Adam. Could Adam have been anything other than Hebrew? Of course not. So if Adam were the first man, wouldn't we all be Hebrew? But we're not all Hebrew, which only goes to prove that at some time in the dateless past, prior to the creation of Adam and Eve, God had created another man and woman who the rest of us (gentiles) are related to. And these people survived the flood of Genesis 1:2. No one will ever know who the first man was—only that he was created by God, not started from pond scum as science has postulated.

So, we have just proven that the Hebrews were a recent addition to the world's population, and that there were no Hebrews before six thousand years ago. So, how could the early church leaders have made such an obvious blunder? And why hasn't something been done to correct church doctrine? Because early church leaders just assumed that the Old Testament applied to all of mankind instead of only the Hebrews.

The reason they don't correct church doctrine now is .. well: "The church is growing and doing well, so why rock the boat over trivialities?" The entire issue has become the elephant in the room that no one wants to acknowledge, and the church continues to preach something that is not true.

Adam was not alone

This revelation is absolutely huge. No, bigger than huge—it sets traditional, old-school religion on its ear and changes everything.

First of all, it proves that there were entire civilizations existing globally, while the Hebrews (God's chosen people) were just coming on the scene somewhere in the Middle East. This idea corresponds with what I had stated earlier, that not everyone perished in God's 2,000-year flood of Genesis 1:2. Scientific proof exists to support this position. If this earlier flood had been the total annihilation of all life, there would have been a glut of remains found dating to around 6,000 BC, followed by a 2,000-year period in which no human remains were found.

This is not the case. This earlier flood was apparently gradual enough to allow considerable populations to somehow survive by escaping to higher elevations, but it is clear that there would have been massive loss of life. However, I have found records of archaeological digs reporting human remains that had been carbon dated throughout the period from 6,000 to 4,000 BC. This would have been during the 2,000-year flood of Genesis 1:2.

This only confirms that there was already life existing on earth when Adam was created. In fact, it is my belief that these earlier people were symbolized by the serpent who convinced Eve to eat of the forbidden fruit. These people would have already been "wise." Besides, using the serpent to symbolize the descendants of Satan's human subjects would have been an ideal way to explain Eve's downfall to Moses, without trying to explain the earlier survivors from the flood of Genesis 1:2.

Not a global flood

This revelation also proves that Noah's flood was not a global flood, as religion has always insisted, but a much smaller, localized flood affecting only the region of God's wrath. Think about it. If Noah's flood had been global in scope, killing all of humanity as well as the entire animal kingdom, and since Noah's family would be repopulating the earth, everyone would be of the same Hebrew nationality. There would be no black, yellow, tan or red-skinned people. Everyone would be Caucasian.

For those who would claim that the distinguishing characteristics of the various nationalities would have been re-established in time after the flood, think again. Common sense would dictate that, if the differences in appearance between nationalities are due to the climates from which they come, it would take hundreds of thousands of years, if not longer, for these people to regain these characteristics. The flood was a mere 4,360 years ago.

Take some time to read and consider the following verses found in Genesis.

Chapter 6:8-13.

⁸But Noah found grace in the eyes of the LORD.

⁹These are the generations of Noah: Noah was a just man and perfect in his generations, and Noah walked with God.

¹⁰And Noah begat three sons, Shem, Ham, and Japheth.

¹¹The earth also was corrupt before God, and the earth was filled with violence.

¹²And God looked upon the earth, and, behold, it was corrupt; for all flesh had corrupted his way upon the earth.

¹³And God said unto Noah, The end of all flesh is come before me; for the earth is filled with violence through them; and, behold, I will destroy them with the earth.

You will notice that throughout the verses quoted above, the point of including the entire earth seems to be implied in the wording selected. However, Moses had no concept of a global earth. His earth didn't expand beyond the area his feet would carry him, or his eyes could see. It should be abundantly clear to even the most critical reader that God was telling Moses the story from his perspective, not from that of 21st century man.

God's new creation targeted

You can also tell from the passages above that God's wrath was directly targeted at Adam's descendants from the re-creation of life in Genesis 1:2 through 1:25, not at any of the surviving descendants of the previous creation, who had re-established themselves in various parts of the globe in the past millenniums. Remember, God was speaking to Moses. This is about the Hebrews and no one else. People of the older creation did not even know or believe in God.

God was specifically targeting His chosen people, the ones whom He had restarted in Eden; the same who had betrayed His love at the hands of Lucifer. These people—His special, super race, to whom He had given an extended life of nearly a thousand years and for whom He had planned eternal salvation—had become evil and wicked, and had turned away from God.

People of the older creation were worshiping the sun, the moon, the rain, or maybe even some old rock they had found. God had no interest in them, no involvement with them. His wrath was aimed directly at the Hebrews and no one else. And, it had only been 1,600 years since He created Adam, so they would have still been in the same region of the earth—the Middle East, the target of God's flood.

Erroneous conclusion from a false assumption

Because of this initial, incorrect belief, there have been other erroneous conclusions drawn. Christianity has traditionally held that Noah's sons, Shem, Ham and Japheth, repopulated the earth after the flood. While this makes for a romantic conclusion to the flood's devastation, it just simply could not be true on so many levels.

First, this would have required them to travel thousands of miles and cross major oceans in order to occupy all the world's continents. Who would the sons' children have found to mate with once they arrived, since all of humanity was gone? And, even if you cross these major hurdles, the time restraints of 4300 years keep this from being a plausible theory. We do, however, know that Noah's sons repopulated the Hebrew world. They were the only ones left to accomplish this. But they just could not have been the originating source of all the other cultures of the world.

An ageless people

Before we continue in determining the validity of the story of Noah's Ark, let's address another possible point of contention for some readers: the claims of the astounding ages achieved by the early Hebrew patriarchs. The Bible reports that Adam lived 930 years. According to scripture, this would mean that Adam actually got to see his great-great-great-great-great grandchild, Lamech, who was Noah's father; this in spite of the fact that men of this period of Biblical history normally didn't marry until the age of 65.

You may be asking yourself, how is this humanly possible? But remember, God wasn't creating just any human. He was creating a special being, using himself as the model. Anything closely approaching a thousand years was a pretty typical life span during the first 1,600 years, from Adam's creation to Noah, who lived to the ripe old age of 950.

But then, something very strange began to happen after the flood waters abated. The longevity of man began to rapidly decline. Less than 400 years after the flood, Abraham only lived to the age of 175. In another 400 years, Moses would only live to see 120 years. And just 400 years after that, David lived a mere 70 years.

So, what was happening to cause this strange phenomenon? There have been many speculations as to the cause of this decline in human longevity. Some have claimed that God initially created man, the earth and its food supply perfectly, but after the flood the earth was completely different than the earth before. Others have suggested that there may have been widespread global changes in climate, composition of the atmosphere, ozone concentration and increasing ultra violet light after the flood. Still others claim that it was due to a deteriorating diet, as well as increasing genetic defaults that caused a rapid decline of the longevity of post-flood humanity.

While all of these claims may sound logical, and certainly they all could have been a contributing factor, I believe there can only be one valid reason for this free-fall decline, only one that embraces sound logic and sensible reasoning. Who were these post-flood survivors going to marry and fornicate with in order to create children?

All Hebrew life was gone, except for Noah and his family. They could have only married the descendants of those who survived the original flood, as described in Genesis 1:2. These people probably had life spans of between thirty and forty years of age. Wouldn't this be the logical outcome of such unions? Not only would life expectancy be dramatically altered, but also, no future generation could be 100 percent Hebrew.

You may be wondering, if people really did live in excess of 900 years, why haven't there been any human remains of this age found in any of the archeological digs? Wouldn't this have made news all over the globe if such remains were found? And I would have to concur, but let me offer this.

It is very likely that there have been a few of these multi-centurion Hebrew ancestors from before the flood unearthed in scientific digs. But quite frankly, how would any anthropologist be able to determine the age of such human remains, if God dialed back the aging process by a factor of ten? A 900-year-old individual may have the appearance of a 90-year-old person. Science can only date a corpse's chronological age by comparing it to the aging process they are familiar with in today's man.

The building of the Ark

In terms of historic exploration, finding Noah's Ark would have to be the most prized and sought-after treasure of all human endeavors. Countless missions have been organized, financed and launched in an effort to prove once and for all that this ancient Biblical occurrence is true. And while there have been some compelling finds providing hope for such an outcome, the Ark itself has never been found.

Mount Ararat, the landing site given in the book of Genesis, is a nearly 17,000-foot elevation mountain in modern-day eastern Turkey. This area has been the primary focus for these excavations. However, many other surrounding areas have been searched as well. The great difficulty lies in piecing together enough useful information, from the sketchy details given in the book of Genesis, to give any real assistance in locating the one find that would make headlines the world over.

We must keep in mind that Moses, not Noah, is telling the story. So obviously, Moses would not have been given all the information required in order to carry out God's command. Consequently, in reading Moses' account of events, we are learning only the bare essentials of what is needed in order to understand the story. So, let's start from what we do know and build from there.

First, God instructed Noah to build the Ark out of "gopher wood". Of course, knowing the type of wood the Ark was constructed from wouldn't be of any assistance in locating it, but it certainly would help authenticate such a find. The problem is, no one really knows what gopher wood is. This seems to be another of those Biblical mysteries that has every expert scratching their head in wonderment.

I believe the most logical explanation is that something went awry in the translation process, going from Hebrew to Greek, and then to English. Some of the more modern versions of the Bible have simply changed the words "gopher wood" into cypress, which isn't a bad guess, according to my research. Cypress has been confirmed as one of the woods which once grew abundantly in ancient Armenia, a country of western Asia, included now in portions of Russia, Turkey and Iran. This is the same area where Mount Ararat is located.

Although cypress is a softwood, it has traditionally been grouped and manufactured with hardwoods, because it grows among hardwoods. But the fact that its pulp is soft would have made forming it into beams possible using the primitive tools of Noah's time. Also, the cypress grows straight and tall, and is a naturally decay-resistant wood, which is reflected in its uses today as a popular choice for building construction, posts, beams, decks and porch flooring. Because of its water tightness, it is also used for shingles, in vats, ship and boat building. Simply put, if the Ark were being replicated today, its builders would certainly specify cypress as the material of choice.

God's dimensions of the Ark are plainly given in Genesis 6:15 as 300 cubits long, 50 cubits wide and 30 cubits high. A cubit is an ancient unit of measure used during Biblical times. While it is anything but an exact unit of measure, literally everyone carried a "cubit-stick" with them. It is the distance from the elbow to the outstretched fingertips, approximately eighteen inches. A "span," another term frequently used, equals half a cubit. Of course, the cubit varies with the length of the individual's forearm doing the measuring, but in using this ancient system of measuring, the Ark would have been roughly 450 feet long, 75 feet wide and 45 feet high.

It would only be natural for one to wonder if there would have been adequate shipbuilding expertise in place during Noah's time to make the building of such a vessel possible. While Biblical historians tell us that Noah's flood occurred in 2348 BC, the Archaeological Institute of America reports that early Egyptians knew how to assemble planks of wood into a ship hull as early as 3000 BC. So, science confirms that the technology existed at least 600 years before Noah is reported to have built the Ark.

Egyptologist David O'Connor of New York University discovered a group of 14 ships entombed at Abydos, Egypt. The oldest of these, a 75-foot vessel dating to 3000 BC, had woven straps used to lash the planks together, with reeds or grass stuffed between the planks to seal the seams. However, the Bible is clear in Genesis 6:14 that pitch, a petroleum-based product with a very thick viscosity, was used to seal the entire hull of the Ark inside and out.

Very likely, wooden pegs known as "treenails" would have been used to connect the planks and other construction members together in the ark, using mortise and tenon joints. This was another technology known to exist in Noah's time. As the treenails became wet, they swelled, which tightened the seal to protect against leaking. Drilling into soft wood with a flint chisel bit, in order to form the hole to receive the treenail, is consistent with tools known to have been in use in the third millennium B.C. Since this was during the Bronze Age in the Middle East, in addition to the bow drill, they would have also had use of both saw and adz, as well as the awe and mallet.

So, having the availability of a suitable building material, the technological skills needed, and the tools required, with adequate time to complete such a project—which Noah's lifespan certainly provided—what else could possibly exist to prevent one from believing the Noah's Ark story? How about the sheer massiveness of the ark? It was, after all, six times the size of the example Egyptian boat found at the Abydos sight . . . the length of one and a half football fields. How could Noah and his three sons have possibly completed such an inconceivable task?

We know that the great pyramids at Giza must have required thousands of people to somehow lift the stones into place. Where did Noah get the help he needed? I have been asking myself this very question for many years. I must have read the story of

Noah's Ark a hundred times, searching for a clue of some kind; something that would finally solve this one perceived flaw that has probably prevented millions of people from truly believing and accepting the story as fact.

To my amazement, I finally discovered it. Why did it take me so long to finally see the answer?

Who helped Noah build the ark?

If you were to open your Bible to the book of Genesis in the Old Testament, and turn to Chapter 7:1, you would see this revealing verse.

Genesis 7

[1]And the LORD said unto Noah, Come thou and all thy house into the ark; for thee have I seen righteous before me in this generation.

The words "and all thy house" appear nowhere else in the entire story, but God is definitely inviting others in addition to Noah's immediate family onto the Ark. When you begin to ponder what life must have been like on 23rd-century B.C. earth, the significance of this statement comes screaming at you with full velocity. Biblical times required an encampment of people living together as an extended family, although not necessarily related. Today we would call this arrangement a commune, but in Biblical times, it was simply called a house. It was a community living together out of necessity, not convenience.

The word house or household appears exactly 1,682 times in the Old Testament alone, so it must have been an integral and necessary element in Biblical living. The reason? You must have others to depend upon in order to merely survive against the elements.

Today, we have become so independent of one another through our modern way of living that if we run out of bread, we just go buy another loaf. But our early predecessors had to first plant, cultivate and water the seed in order to raise the wheat, then harvest and mill the wheat into flour, before they could even mix the batter. Then they would have needed to cut the wood to heat the hearth so they could bake the bread. In contrast, all we need today is a set of car keys and some pocket change.

To confirm the fact that this was the typical living arrangements in Biblical times, all one needs to do is read Genesis 36:6.

Genesis 36

⁶And Esau took his wives, and his sons, and his daughters, and all the persons of his house, and his cattle, and all his beasts, and all his substance, which he had got in the land of Canaan; and went into the country from the face of his brother Jacob.

In Noah's case, "all the persons of his house" could have included field workers to produce the grains, and orchard keepers to tend the tree crops, such as olives, figs, dates, nuts and apples. Then there would have been the fruit dryers and processors to dry and pack the fruit for storage and later use. Also, there would have definitely been vineyard workers to grow and tend the grapes, and then make the wine.

Noah and his wife would have also needed toolmakers, pottery workers, cooks, clothing makers, and carpenters as well as shepherds to protect the flocks and cattlemen to manage the herd. Then there would need to be butchers to prepare the meat for cooking. Many of these workers probably had families of their own. I can visualize there being as many as fifty or so members of Noah's household, and I'm sure they would have been more than happy to help build the ark and take care of all the animals, once they were underway, just for the chance to avoid certain death if they remained behind.

Now, I must ask you before we continue on this quest for truth: Isn't this Bible story beginning to sound less like a fable, and more like an actual event? Next, let's consider what Noah's Ark may have looked like if you were actually there.

Imagining the Design of Noah's Ark

Although the curved hull design of today's sailing vessels was known to exist at least 600 years prior to the building of the ark, this hull design would not have been necessary for the ark, since it would not be sailing anywhere. The only requirement would have been to stay afloat and dry for about a year during God's deluge. So, its hull design probably resembled that of a johnboat or barge, with a flat bottom and perpendicular sides. With the use of mortise and tenon joining methods, which is known to have existed during Noah's time, any construction configuration would have been possible.

According to Genesis 6:16, there were three stories below the main deck, making a total of four decks. The main entryway to the ark would have been in the side of the ark below the main deck. Therefore, the decks would have been about 15 feet apart, a total of 45 feet in all.

In the diagram below, you will notice a single "rampcase" amidships in order to easily move the animals between decks. A rampcase has sloped ramps instead of stairs; thus the name. You may imagine it similar to the way a parking garage is laid out. One complete trip around the rampcase would take you to the next level, either higher or lower, depending on which way you were going.

In the center of the rampcase, and in 60-foot increments from bow to stern, there would have been large vents measuring 15 feet by 15 feet, open on all sides and at every level, to release the methane gas created by animal excrement from below decks. The perimeter of the openings in the main deck would have been built up about a foot high to prevent rain from vent. These would be covered by large four-sided tents, fashioned from woven strips of

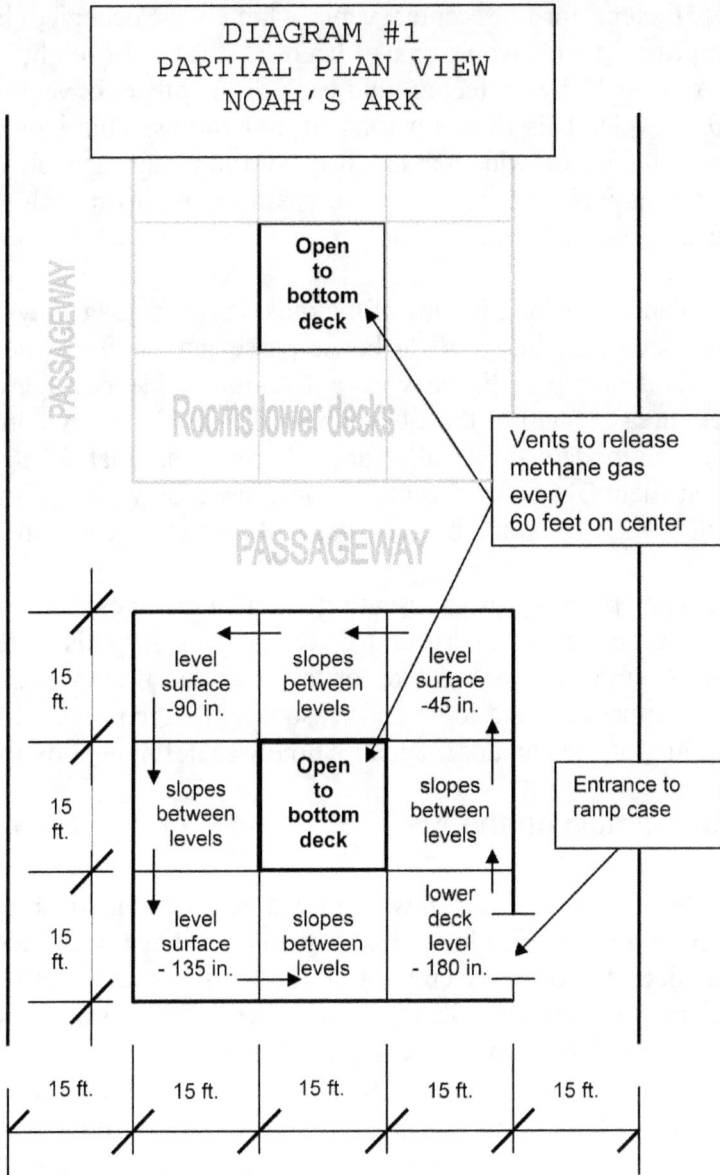

DIAGRAM #1
PARTIAL PLAN VIEW
NOAH'S ARK

Open to bottom deck

PASSAGEWAY

Rooms lower decks

PASSAGEWAY

Vents to release methane gas every 60 feet on center

15 ft.	level surface -90 in.	slopes between levels	level surface -45 in.
15 ft.	slopes between levels	Open to bottom deck	slopes between levels
15 ft.	level surface - 135 in.	slopes between levels	lower deck level - 180 in.

Entrance to ramp case

15 ft. | 15 ft. | 15 ft. | 15 ft. | 15 ft.

Width of Ark 75 Feet

material sewn together and coated with pitch. These would have been removed after the rains abated.

There would have been one of these vent openings every 60 feet, making a total of seven over the entire 450-foot length of the ark. Except for the center vent, which was occupied by the rampcase, there would have been a cluster of eight rooms measuring 15 by 15 feet around each of the other six vents on all three levels. This makes a total of 144 rooms. This would also leave a 15-foot wide passageway surrounding each cluster of rooms, as well as a 15-foot wide passageway down each side of the ark on every deck.

Ten-foot wide sections of the outside two passageways could have been partitioned off to house passengers and provide room for food storage, still leaving a five-foot wide passageway in these areas. If half of the 144 rooms were divided in two, in order to accommodate the smaller animals or fowls, that would have given them 216 rooms in all. If there were only two animals in each room, that would be a total of 430 critters saved on the ark.

The reason I'm presenting you with these plans and calculations is not to try and convince you that this is exactly what Noah's Ark looked like, or to prove that there was such an ark. But using sound logic and reasoning, I have proven that the building of such an ark in Noah's time was definitely possible.

The Animals on the Ark

Since we know this was only a regional flood and not a global flood, we also know that it would not have been necessary to collect two of every creature in the world to save on the ark. I will remind you again that Moses, to whom God was telling this story, would not have understood the concept of a global flood. To him, all the animals of the earth would have been all the animals of his immediate geographic area in the Middle East. He would not have been aware of every animal in the animal kingdom.

In Genesis 7:2-3, we read:
²Of every clean beast thou shalt take to thee by sevens, the male and his female: and of beasts that are not clean by two, the male and his female.

³Of fowls also of the air by sevens, the male and the female; to keep seed alive upon the face of all the earth.

In these verses, it is clear that Noah is being made responsible for gathering only domestic animals and birds, such as cattle, sheep, horses, camels, chickens, and geese. The "clean beasts" collected by sevens will be used for food, and the unclean collected by twos for saving on the ark.

And then in Genesis 6:20 below, we can understand that any of the wild animals would be delivered to Noah by God for him to keep alive. Notice God says "two of every sort shall come unto thee." He says nothing about gathering them, but he will require Noah to gather the food "for thee and for them" in verse 21.

²⁰Of fowls after their kind, and of cattle after their kind, of every creeping thing of the earth after his kind, two of every sort shall come unto thee, to keep them alive.

²¹And take thou unto thee of all food that is eaten, and thou shalt gather it to thee; and it shall be for food for thee, and for them.

So, it would not have been Noah's job to play animal catcher. He was busy building an ark. But if God is capable of creating a universe, along with all its inhabitants, would it be too much to believe that he could also convince a few animals to show up at Noah's place for launch week?

Anyone looking for geological proof of Noah's flood won't find it, unless someone eventually finds the ark, because raining for forty days and forty nights is not nearly long enough to leave any clues in the earth's strata. So, if you are looking for hard and fast proof that there really was a Noah's Ark, I'm afraid it eventually boils down to having faith in what the Bible says. But on the upside, I've never found it to be wrong yet.

Chart #5
40,000-Year Timeline
Of Human Existence

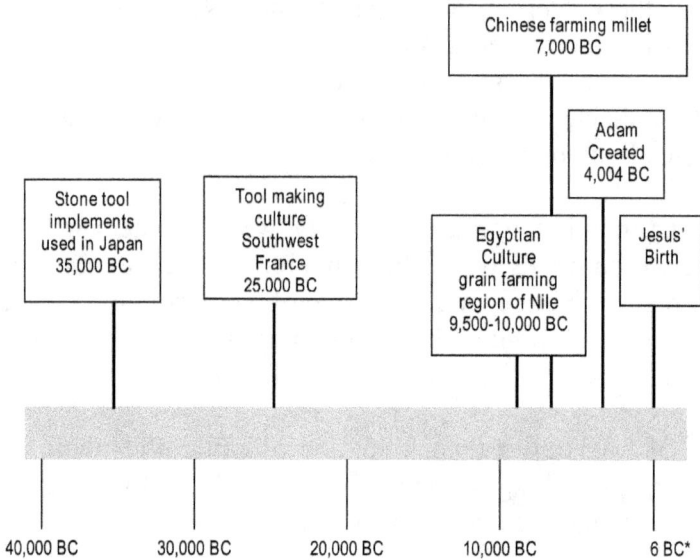

				Chinese farming millet 7,000 BC

Adam Created 4,004 BC

Stone tool implements used in Japan 35,000 BC	Tool making culture Southwest France 25.000 BC		Egyptian Culture grain farming region of Nile 9,500-10,000 BC	Jesus' Birth

40,000 BC   30,000 BC   20,000 BC   10,000 BC   6 BC*

You can see by this 40,000-year timeline, as human cultures go, the Hebrews were latecomers to the earth's stage. Many human cultures are known to have existed thousands of years before Adam.
This is still more scientific evidence supporting the existence of a divine creator since their findings agree with the bible when properly interpreted.

Chapter 8

The Battlefield

Science and religion had to ultimately clash. It was only a matter of time. And one would think that the field of archaeology would be the likely candidate to lead the charge against God, in an attempt to unseat Him as being creator of all living things. By definition, archaeology is the study of human antiquities. But since human remains must be carbon dated to determine age by testing the surrounding strata, paleontology—the study of fossils—has probably become the prominent aggressor in this war.

As with the rest of the scientific encampment, paleontology denies the existence of a Creator God, which is diametrically opposed to the view that religion holds. This lack of recognition only feeds the flames of war. And these battle-ready warriors have other compatriots fighting in the trenches with them: anthropology (the study of humanity), sociology (the study of society), biology (the study of natural science), genomics (the study of genomes), and evolutionary biology (the study of origin of species), just to name a few. At some point, this army of scientists became bored with simply studying and decided to conquer. Sometimes too much knowledge can have an intoxicating effect.

Of course, the other encampment has done its share of stoking the fires of discontent. For thousands of years, religion has put its Creator God above all things earthly, demanding that all of humanity pay homage to a deity that you cannot see, feel or touch. A God who is portrayed as loving, caring and compassionate, but who allows bad things to happen to good people, and one who seems to regularly turn his back on evil. If He isn't there when you need Him, maybe He isn't there at all. How can religion blame science for trying to fill what they interpret to be a gaping void in the greatest of human yearnings . . . the answer?

As with any war, someone had to fire the first shot. That first shot came as a lawsuit from the religion encampment. Suddenly, the battleground became a court of law, where the "trial of the century" was played out in a sweltering courtroom in the little town of Dayton, Tennessee, in the summer of 1925. A high school science teacher by the name of John Thomas Scopes was put on trial by the state of Tennessee for violating Tennessee's Butler Act, which made it illegal for evolution to be taught in any state-funded institution. The Butler Act had just been passed in both the state house, on January 28th, and in the state senate on March 13th of that same year.

This legal action couldn't have been more publicized—in part because two of the most famous big-name lawyers in the country were going to be arguing the case, but also because God's law was being tested in a human courtroom. The newspapers had a field day in reporting on what became known as the "Scopes Monkey Trial."

Arguing for the prosecution was William Jennings Bryan, three-time presidential candidate on the Democratic ticket. Clarence Darrow, the famed defense attorney, represented Scopes. Both of these lawyers were known to be flamboyant showmen with egos to match, and newspaper reporters from across the country swamped this sleepy little Bible-belt town to cover a story that was guaranteed to be monumental, in terms of setting a legal precedent on the controversial topic of religion versus science.

The teaching of evolution had begun to gain acceptance with mainstream America, and fundamentalists saw the state courts as a way to turn back the tide toward creationism. With the reputations of two notable lawyers at stake, the proceedings became a contest of wits.

Seven days into the trial, Clarence Darrow, in an almost unprecedented move, called his adversary, William Jennings Bryan, to the stand with a scheme in mind. His plan was to question him about the stories of the miracles found in the Bible, hoping to draw responses too incredible to be believed. The prosecution accepted, with the understanding that Bryan could then question Darrow on the stand.

The day was exceptionally hot, so Judge Raulston adjourned court to be held on the courthouse lawn. With an atmosphere reminiscent of a Fourth of July outing, Darrow began his questioning. His prepared topic on Adam and Eve included questions like, was Eve actually created from Adam's rib? Each question drew a snide response that sidestepped the anticipated incriminating result. Then Darrow would counter with an equally derogatory remark. When asked where Cain got his wife, Bryan retorted, "I would have the agnostics to hunt for her."

The session grew more hostile with each exchange. Finally, Darrow snapped, "You insult every man of science and learning in the world because he does not believe in your fool religion."

In response, Bryan declared, "The reason I am answering is not for the benefit of the superior court. It is to keep these gentlemen from saying I was afraid to meet them and let them question me, and I want the Christian world to know that any atheist, agnostic, unbeliever, can question me anytime as to my belief in God, and I will answer him."

Then the deputy prosecutor objected, demanding to know the legal purpose of Darrow's questions. Fearing the effect the questions were having on the jury, Bryan snapped, "The purpose is to cast ridicule on everybody who believes in the Bible."

Then Darrow spewed in response, "We have the purpose of preventing bigots and ignoramuses from controlling the education of the United States."

After a few more exchanges, Judge Raulston, seeing that nothing was being accomplished, ruled the entire exchange irrelevant, banged his gavel and adjourned the court.

Realizing they were headed for appeal, Darrow offered no closing statement, which under Tennessee law also banned the prosecution from final summation.

It took the jury only nine minutes to bring back a guilty verdict for Scopes. Judge Raulston fined the defendant $100, which, adjusted for inflation, would be the equivalent of $1,325 today.

The case was appealed to the Tennessee Supreme Court with four claims:

1. The term "evolution" in the state statute was too broad.
2. The statute violated Scopes' constitutional rights to free speech. (The state reserved the right as employer to determine what was taught.)
3. The Butler Act was in violation of the Tennessee State Constitution.
4. The Tennessee State Constitution forbid the establishing of a state religion.

The Supreme Court rejected all claims, but they did overturn the conviction on a technicality—that the jury should have set the fine, not the judge. Tennessee judges could not exceed $50 as a fine.

Then Justice Green added a totally unexpected recommendation, that any further attempts to prosecute would fall under the jurisdiction of the state attorney general's office. Attorney General L.D. Smith immediately announced that he would not seek a retrial.

While it was unclear whether it was science or religion that actually won in this initial battle, it was undoubtedly science that won the war. In 1968, the Supreme Court of the United States ruled in Epperson vs. Arkansas 393 U.S. 97 (1968) This decision overrode the states' constitutions that banned evolution from being taught in schools, and found the teaching of creationism to be in violation of the establishment clause of the First Amendment.

Further attempts to allow the teaching of creationism alongside evolution have failed. Today, it is illegal to teach creationism in any public school.

Did you know . . .

. . .God denies that He created the earth using evolution? Read about it in the next chapter.

Chapter 9

Evolution vs. Creation

Even though the battleground between science and religion became the courtroom in 1925, the enmity between the two actually began in the 19th century during the Victorian Era, with Darwin's theory of evolution. Ironically, this theory itself seems to have evolved with time. That is to say, what began as Darwin's original *Theory of the Origin of Species* stands today as something quite different than what Darwin originally proposed 150 years ago. This is especially true with mankind's historic beginnings. It seems that every time ancient human remains are unearthed, man's ancestry changes to fit the find.

Charles Darwin's Theory of Evolution is one that has been introduced, packaged, promoted, sold, defended, and ultimately accepted as truth by millions of people. But there remains just one small detail that seems to have been lost. Until a theory is proven, it is still merely speculation. Even the name, whether intentional or not, exhibits real marketing genius. According to Webster, the meaning of evolution's root word, "evolve," is to develop slowly. The word has nothing to do with changing, as Darwin's theory proposes.

So, a more accurate label for it might be the Theory of Transformation, which means to change forms. However, since it is only human nature to resist change, it is doubtful that his theory would have caught on had the name implied changing rather than developing. Developing sounds so natural, progressive and positive. Changing can be so uncertain and fearful.

I've always wondered too, exactly who is it that decides what material will be presented in elementary textbooks for young minds to absorb? I remember seeing some real neat pictures of Cro-Magnon man and Neanderthal man in my fifth grade science book, and I'm nearly a decade past retirement age. What I don't remember seeing is the word "theory."

Even though it was probably there, what impressed me most at the age of ten were the pictures of very realistic looking, prehistoric humans in my textbook. And my teacher was one whom I looked up to and respected, so I accepted anything she said as absolute truth. I'm sure these same pictures would have been in her science book when she was a fifth grader. After all, this process for teaching evolution has been in place for nearly five generations.

I think the question must ultimately be asked: If Darwin's *Theory on the Origin of Species* turns out to be wrong, how do you un-teach hundreds of millions of believers? How do you reverse an erroneous belief that stretches back to include our great-great-grandparents? If evolution is eventually proven to be a flawed theory, what administrative body in charge of the scientific community exists, and has both the power and the authority, to reverse a decades-long belief? Or maybe a better question would be: Even if they could reverse their position, would they?

I think you can get your answer to this question by looking at the religious leaders of today, who continue to sell their flawed theory on creation even though it is not true. The answer is. . . no one. No one would step forward to claim responsibility for causing humanity to follow the Pied Piper. I'm not naive enough to believe that my humble writings are going to change the unchangeable, but I must reveal my findings. To do otherwise is simply unthinkable.

Darwin's theory

Charles Darwin's *Theory on the Origin of Species*, which has been promoted as "the theory of evolution," has come to be based on two assumptions. The first is a very sound premise, which Darwin called natural selection. This is now realized to be one of God's natural laws of the universe, its effects every bit as real and undeniable as the Law of Gravity. In layman's terms, the Law of Natural Selection allows for gradual shifts to occur in the characteristics of a species, in order to perpetuate and protect it from extinction by its predators or any other natural cause in nature. It works by allowing cells with sound characteristics to live, and those with flawed or undesirable traits to die off.

Perhaps a simpler way to think of it is best illustrated in the axiom "survival of the fittest." When it was realized that this law only worked within a species, and did not promote change into a more advanced species, proponents of Darwin's theory added a second assumption called the Random Mutation Theory.

When a cell dies, it clones itself into an exact replacement cell. This new cell has the same DNA and other characteristics as the initial cell. This continual turning over of cells is the definition of life.

The Theory of Random Mutation states that occasionally a cell misfires and creates a replacement cell that has corrupted data, and while almost all mutated cells cannot survive, they believed a rare few turned out to be beneficial cells, which eventually became more complex life forms.

So, the Theory of Evolution now combines the Theory of Random Mutation with the Law of Natural Selection, over a period of billions of years, to create all the plant and animal life forms we have today. Yes, you read that correctly. Bad can actually turn out to be good. Flawed can be flawless. Wrong can be right, and misfires can correct themselves to be beneficial.

Let me give you a modern day example of the Random Mutation Theory at work. You wreck your 1987 Ford Fairlane. The tow truck comes and tows it to the garage, where you have the necessary repairs made. A few weeks pass before they call to let you know that your car is ready. You go to the garage, get into your repaired Lamborghini (Italian sports car) and drive it home. I realize this comparison is a little flippant, and I don't know what your experience has been—but any time I have had corrupted data, the outcome is never good.

The assumption is that life on earth began billions of years ago in the waters of earth, with a single-cell organism. After billions of years, through the process of evolution as described above and a whole lot of luck, this organism changed into a chordate (eel-like creature). Darwinians believe this process continued throughout time, causing the chordate to became a tetrapod (fish), then a reptile (lizard), then a mammal (rodent), then a primate (monkey), then a Homididae (Ape), and finally Homosapien, man's direct ancestor. Now, think about it. Could millions of complex plant and animal species be the result of a few happy accidents from mutated cells left to restructure themselves over billions of years?

I consider myself to be a person of only average intelligence, but even I have come up with five other major flaws in Darwin's revised theory—and these are in addition to the Random Mutation Theory, which I find to be totally unbelievable.

Flaw #1 – Any change must always be supported by a reason. The motive for change, which occurs through natural selection, is always preservation from extinction. Evolving from a lower life form to a higher life form would lack this motive, and in fact lack any reason for change whatsoever.

Flaw #2 – The listing of life forms above represents earth's entire food chain. To evolve from the bottom up would destroy the very food that the more advanced life forms would depend on, therefore causing all life to become extinct. This is true because as particular species evolve to a more complex species, it would leave vacant the species it once was.

Flaw #3 – Either all life forms must evolve together, or not at all. There is no explainable reason why only some forms would evolve, and others would remain to provide the food for those evolving. Science cannot have it both ways.

Flaw #4 – Even if there were some kind of partial evolutional change going on, we would see evidence of multiple in-between stages of life species. This "missing link" has never been found between any of the established life forms. This lack of in-between stages of life proves that there was no gradual change occurring, as Darwin postulated in his theory. Finding just one example of a changing species would at least give evolution the hope of being real.

Flaw #5 – Darwin's theory asserts that all living matter, whether plant or animal, began from single-cell organisms in the sea. The example above represents man's evolutional pathway into the present. Birds and other plants and animals would each have a different pathway. However, this leaves an even greater puzzle to solve. How can four separate cells, identical in every respect, produce a man, an alligator, an eagle, and a rose after billions of years of evolution? The real irony is this: Ask any scientist why he doesn't believe in a creator God, and he will probably reply, "It's just too far-fetched to believe." Oddly, he doesn't regard Darwin's theory the same way.

The Darwin Fraternity

I've seen this situation in almost every walk of life, and in most organizations as well. It's one of those human conditions that stagnates growth and brings progress to a grinding halt. It corrupts like a cancer. I've seen it in clubs, companies, religion, politics, government, the military, and in practically every human grouping you can imagine. It's also known as the "good ol' boys' network." And sadly, it's alive and well in the scientific community.

Anytime someone comes along with a different view of an established concept, those in the group who like to control things start the rejection process, which includes every trick in the book in order to kill the beast. "The Fraternity" is responsible for more stagnation than all other causes combined. It operates with a Klan mentality.

Case in point:

Geneticist Dr. Barbara McClintock was one of the brightest minds in her field and was decades ahead of her time. She discovered cellular engineering and won the Nobel Prize for this discovery in 1983. Unfortunately, her discovery butted heads with Darwin's theory of evolution. The reception of her research was greeted with skepticism and even hostility from her peers, and she began to feel alienated from the scientific mainstream. In 1983, she stopped publishing accounts of her research. Today, her findings are not even being taught in most biology classes.

This is just another example of the power and control that the Darwin Fraternity wields. But I feel that if a theory can't be proven in more than 150 years, it should be time to repeal its theory status, return to the drawing board, and begin to consider some other possibilities.

The Theory of Creation

God doesn't waste any time making his case. The first verse in the first chapter of the first book of the Bible is: "In the beginning God created heaven and earth." These were the words written by Moses, who began penning the first five books of the Old Testament nearly 3,700 years ago.

In Genesis 1, verse 2, which begins billions of years after the creation in verse 1, God describes the flooded condition of the earth as it was before He gave His verbal commands to restore it. Scientific findings have confirmed this flood, which existed on the earth at precisely the time the Bible records it.

In verses 3-19, God describes the first four phases (days) of restoring the earth to a habitable state, which also includes some references to the original creation. These phases were necessary in order to prepare for the creation of life on earth.

(NOTE: This can be somewhat confusing until you realize that God is combining the story of the original creation with the story of restoring the earth to prepare for the creation of Adam and Eve.)

In the fifth phase, God created fish and then fowl in verses 20 and 21. In verses 24 and 25, He created all the land animals. It is clear that God is setting up the world's food supply from the bottom up—first in the sea, then in the air, and finally on the ground. Science has confirmed this sequential order of existence by dating the fossils of early life forms. (*See the chart at the end of this chapter.*)

In the sixth and final phase, verses 26 through 31, God creates Adam and Eve, the first Hebrew couple, and describes their role on earth.

Notice that God tells his story in a way that Moses would understand, using the word "day" instead of confusing him with time expressed in billions of years, which Moses couldn't have possibly comprehended.

Now seriously, given the limited knowledge of Moses, as opposed to the advanced knowledge you posses by living in the 21st century, could you have told Moses the story of creation any better than God did? And while you're contemplating things, consider this: How could an ancient manuscript, written by one who knows nothing of science, so closely resemble Darwin's theory, only free of the six flaws explained above?

Oh, I know the one perceived flaw that may still be lingering in your mind: "But is there really a creator being capable of performing such miraculous feats?" And to that thought, let me respond: Can you really think of a more reasonable scenario?

Evolution, Biblically denied

God clearly denies in scripture the use of evolution in creating the various life forms. The following two verses from the creation story will confirm this. Exactly sixteen times in scripture, God makes it a special point to explain that each new creation is either created *after his kind, after its kind,* or *after their kind.*

Never is it stated that something was created after an earlier or previous kind, as would be the case with evolution.

Genesis 1:21 And God created great whales, and every living creature that moveth, which the waters brought forth abundantly, *after their kind,* and every winged fowl *after his kind:* and God saw that it was good.

Genesis 1:25 And God made the beast of the earth *after his kind,* and cattle *after his kind,* and everything that creepeth upon the earth *after his kind:* And God saw that it was good.

In the end, you will believe what your convictions dictate. It's not my intent to overturn what is in your heart. I can only present the facts as I have discovered them, and let you judge their value in your own life.

Did you know . . .

. . .throughout the past 150 years, there have been many different versions of the Theory of Evolution? This is explained in the next two chapters.

Chart #6
Earth's last 2 billion year timeline
(Evolution or Creation?)

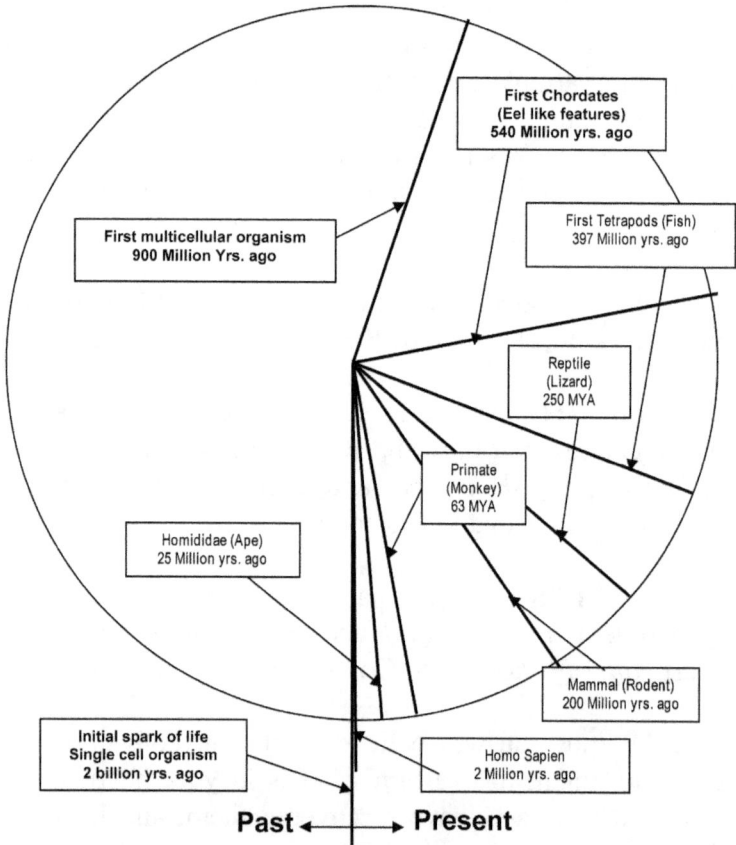

First Chordates
(Eel like features)
540 Million yrs. ago

First Tetrapods (Fish)
397 Million yrs. ago

First multicellular organism
900 Million Yrs. ago

Reptile
(Lizard)
250 MYA

Primate
(Monkey)
63 MYA

Homididae (Ape)
25 Million yrs. ago

Mammal (Rodent)
200 Million yrs. ago

Initial spark of life
Single cell organism
2 billion yrs. ago

Homo Sapien
2 Million yrs. ago

Past ◄─┼─► Present

Do you believe all this really happened
over billions of years as a result of a fluke? (Evolution)
or has God created the world's food chain
from the bottom up? (Creation)

Chapter 10

The Changing Theory of Man's Evolution

In the previous chapter, I have covered the subject of Darwin's Theory of Evolution in a general sense, from the beginning of single-cell life in the waters of earth two billion years ago, to the millions of complex plant and animal species that we see today. However, this thesis concentrated on tracking only man's evolutional pathway into the present, not the pathways of all plant and animal life on earth. I also covered the theory of creation during the same period of time.

The next two chapters will deal more directly with what paleontology considers to be mankind's emergence onto the world stage, roughly six million years ago, and carry it forward into the present. I must point out that the views from the theory of creation's perspective is not that of traditional beliefs, but based rather on what I believe to be the only reasonable interpretation of the Old Testament text.

It is important to know that the world of science has developed a unique vernacular. For example, when you see the letter designation BP following a number, it stands for "before present." So, 1000 BP is the scientific way of saying "one thousand years ago." Time may also be stated as 10 MYA, meaning "ten million years ago." In the interest of forest conservation, however, a layman's translation will be provided for each of their encrypted statements.

If you can't convince `em, confuse `em

What began as a simple premise in 1859 with the writing of Darwin's *The Origin of Species* has "evolved" into a hodgepodge of interwoven theories and hocus-pocus—so confusing and ever-changing, it seems that even the most learned individual would want to bang their head against the wall repeatedly in order to receive some small degree of relief from their intense frustration. If you have studied and tried to understand the latest version of the theory of evolution, you must know the feeling.

I am going to give you a simple chronology of events in four phases, in order to untangle what I fully believe to be a web of confusion, which has been spun over the past 150 years in order to shore up a flawed belief that all life originated from a single cell in the sea two billion years ago. It is not my intention to paint the entire scientific establishment with a deceitful brush. I do think they have found creationism so impossible to believe that they are able to see logic where no logic exists. I think the Darwinist movement has defended their position so long, that to back down now would certainly destroy the credibility of the entire scientific establishment. So they just keep going, applying layer after layer in an attempt to permanently forestall the inevitable.

Phase I – Darwin's discovery of natural selection was undoubtedly the most brilliant scientific discovery of the nineteenth century. His documentation proved that a species would actually improve its own ability to survive by making cellular adjustments, thus changing its physical characteristics. This discovery in and of itself should have been convincing evidence of intelligent design, but this was ignored instead. Darwin was convinced that his theory connected all plant and animal life back to the single-cell organism of two billion years ago.

Phase II – Late in the 19th century, when science finally discovered that changes brought about by natural selection did not cause a change of species, the Theory of Random Mutation was introduced. Working in tandem with natural selection, they believed that a mutated cell, strengthened by the function of natural selection, would give the boost needed to "jump the gorge" into a new and different species. In time, they realized they were no closer to proving evolution than Darwin had been— yet they marched on undeterred.

Phase III – Early in the 20th century, science began broadening their view of evolution to include not only Darwin's Theory of Natural Selection and the Theory of Random Mutation, but also Gregor Mendel's basic understanding of genetic inheritance. Gregor Mendel was an Austrian monk who discovered the laws of genetics in the 1860s, when he crossed pea plants to see how traits were passed from one generation to the next. During this period, science also began to include new theories from field biologists, population geneticists, and later by molecular biologists. This broader view is known as "Synthetic Evolution."

Phase IV – Today, scientific research has concentrated on what is known as "Evolutionary Developmental Biology," or "EVO-DEVO" for short. It focuses on genetic changes that alter embryonic development. Their goal: To find new features in species lines, which would be an indicator of evolution.

Needless to say, nearly everyone today believes in evolution. Five generations have been educated to accept it as fact, even though it has never been proven. Even most of those who believe in creationism believe God created using evolution.

Above, I have covered what I consider to be the four "evolutionary" phases of the Theory of Evolution. Now, let's put them under the microscope to see how they hold up under the scrutiny of sound logic.

Re-writing the Theory

Although Gregor Mendel had discovered the Laws of Genetics in the 1860s, it wasn't until the turn of the century that science realized this to be the third factor affecting evolution. Biological science finally redefined evolution to be: "The sum total of the genetically inherited changes to the members of a population's gene pool." In other words, if the varieties of a particular gene (two alleles, pronounced ah-lee-lees) increased from one generation to the next, within the entire population as a whole, then evolution had occurred.

This definition was adopted in 1908 as a result of work done by Godfrey Hardy, an English mathematician, and Wilhelm Weinberg, a German physician. They determined that evolution to a population was occurring all the time, unless these seven conditions existed:

1. Mutation is not occurring
2. Natural selection is not occurring
3. The population is infinitely large
4. All members of the population breed
5. All mating is totally random
6. Everyone produces the same number of offspring
7. There is no migration in or out of the population

Since the likelihood of all seven conditions existing would be impossible, Hardy and Weinberg felt that evolution was inevitable and occurring continually.

A Crash Course in Genetics

It would be beneficial to pause here and become familiarized with the basics in the Laws of Genetics, since the shift in the theory of evolution has evolved in that direction.

The genetic makeup of a living organism is called its *genome*. In the nucleus of every human body cell are 46 chromosomes. A chromosome is a strand of DNA, along with associated proteins that carry the organism's hereditary information. This information is carried in the genes, which are attached to the chromosome.

It is the position (locus) of the gene on the chromosome that determines the character trait, and not the gene itself. *(More about chromosomes later.)*

Each gene is comprised of two alleles (ah-lee-lees). One allele is dominant for a trait; the other is recessive. The dominant allele is represented by an upper case letter, the recessive allele by a lower case letter. These two letters, one for each allele, make up the *genotype* of an organism. The physical appearance of a trait in an organism is called the *phenotype*.

For example, if we were considering color as a trait in mice, gray might be dominant, represented by the letter "G," and white would be recessive, represented by the letter "g". Both alleles are forms of the same gene, so we use the same letter for both.

Our possible genotypes and phenotypes would look like this:

Symbol	*2. Genotype Name*	*1. Phenotype*
GG	Homozygous *(pure)* dominant	gray mouse
Gg	Heterozygous *(hybrid)*	gray mouse
gg	Homozygous *(pure)* recessive	white mouse

Notice that the letters under the symbol column represents two alleles of the same gene, not two genes. Also, in the second line, when dominant and recessive traits are crossed, the result is always dominant. In the third line, it requires two recessive alleles to produce a recessive trait.

Reginald Punnett, who studied Mendel's theories, developed the Punnett Square below, used to determine the possible genotypes of an offspring. Of course, during Mendel's time, genes had not been discovered yet.

Here's how it works:

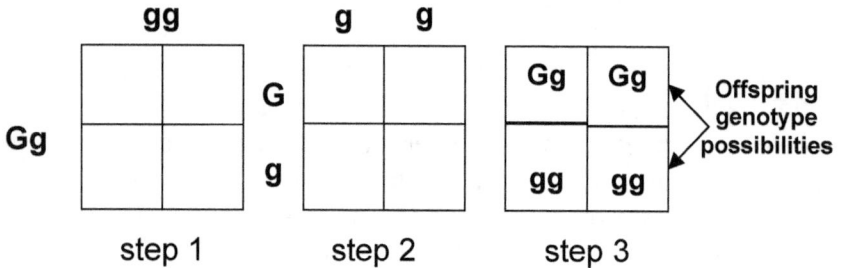

Step 1. Place the indicator for one parent's genotype at the center of the left border, and the other parent's indicator in the center at the top border.

Step 2. Separate the alleles, as indicated in the second step.

Step 3. Carry over the indicators from the left side and carry down the top indicators,
so that you have two alleles (one gene) in each box.

To summarize the results of the possible genotypes
and phenotypes, two of the four boxes have "Gg," which is gray. By dividing the number of boxes with "Gg" in them (2) by the total number of boxes (4), we find the possibility is that 50% of the offspring will be gray mice. (2 / 4 = 50%) Using the same method, we can see that there is a possibility that 50% of them will be white.

Normal body cells have 46 chromosomes each, and when body cells divide and replicate themselves (called mitosis), the new body cells also have 46 chromosomes. However, when the sex cells (sperm and egg) divide in half during meiosis (sex cell replication), when the sperm and egg are fused at fertilization, the new cell (zygote) will receive 23 chromosomes from each parent, giving it the 46 chromosomes it needs. In the Punnett Square example above, this is why the genotypes are split in step 2.

We have seen how the Punnett Square can be used to determine the possible changes between generations in a single case example. However, this wouldn't work in a population or gene pool. Therefore, Hardy and Weinberg came up with an equation that could track these changes from one generation to the next.

$$(p^2 + 2pq + q^2 = 1)$$

Where p = frequency of dominant allele, and where q = frequency of the recessive allele, which control the same trait.

Unfortunately, unless you excel in math, this probably won't make a lot of sense to you. So let me summarize in plain English: Biologists now believe that evolution is simply a change in frequencies of alleles in the gene pool of a population. For instance, take an example of a trait with a gene of two alleles, B (dominant) and b (recessive). If the parent generation has 92% B and 8% b (100%), and their offspring collectively have 90% B and 10% b (100%), evolution has occurred between the generations.

Biologists also believe that, while gene pool frequencies are normally stable, they can change when affected by other evolutionary mechanisms. (See chart below)

Mechanisms that change the gene pool's frequency within a population	**+**	The effects of Natural Selection	**=**	**Results**

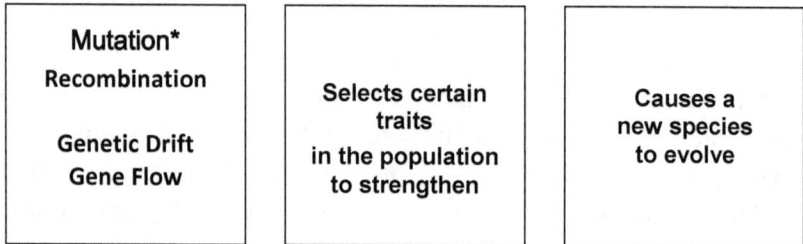

Mutation* Recombination		
Mutation* Recombination **Genetic Drift** **Gene Flow**	**Selects certain traits** in the population to strengthen	**Causes a new species to evolve**

* While mutations neither increase or decrease allele frequencies in a gene pool within a population, science believes it is one of the greatest possibilities to be the origin of new species.

Scientific Claims About the Effects of Mutation and Natural Selection on Evolution

There are two basic kinds of mutation. The most common occurs during DNA duplication in cell division. This is because millions of mechanical operations must be precisely completed in order for new DNA molecules to be created, so the possibility of misfires are more common during this process.

A second form of mutation, called genetic recombination, is far less common. This is where a small amount of genetic material, and less occasionally, DNA, can be transferred by viruses between organisms. It is the belief of today's biologists that when these mutations occur, natural selection can select in favor of the mutation and lead to evolution.

Debunking Mutation as a Cause of Evolution

It's amazing to me what biologists can come up with in order to try and justify evolution. Mutation is not something that happens to species of an entire population. It occurs only rarely in single cells during replication. On the other hand, natural selection only works on an entire species population to protect it from predators or natural changes in its environment. It does this by selecting only those traits capable of combating the problem, and strengthening them generation after generation.

There is no evidence to show that this mechanism would protect a mutated cell's flawed trait and accelerate it. Besides, it takes years, and sometimes decades, for natural selection to begin selecting traits in a species. The life span of a mutated cell is normally shortened because of its fragile state. It is, after all, a misfire. And even if natural selection could select a mutant flaw and accelerate it, wouldn't this make it more flawed? Wouldn't this be the equivalent of changing a species in the wrong direction and propelling it into extinction?

Now, think about it. Does it really make sense that as perfectly as the replication process of cells work, natural selection would mistake a mutant cell for a healthy one, and then select for its flaws to establish a new species? In order for this to happen, two separate mechanisms of nature—cell replication and natural selection—would both have to fail. And not only would they have to fail, they would also have to fail simultaneously. The chances of this happening would have to be 1 zillion[20] to 1.

Debunking the Scientific Claims About the Effects of Genetic Drift and Natural Selection on Evolution

Besides mutation, genetic drift is the other suspected cause of evolution. You'll recall that Godfrey Hardey and Wilhelm Weinberg came up with a "new" definition for the process of evolution, which was adopted in 1908. "If the varieties of a particular gene _increased_ from one generation to the next, within the entire population as a whole, then evolution had occurred." But genetic drift is the "Random _fluctuation_ in the frequency of a particular gene in a small, isolated population."

The key word here is fluctuation. This means that these varieties of a particular gene can *decrease* as well as increase. So, how does one measure increase? Wouldn't it be the difference between the two? And if a number of these varieties decrease more than increase, does this reverse evolution?

Genetic drift is a natural, cyclical event. But, giving science the benefit of the doubt, suppose there were a net increase. How would an increase in the varieties of a particular gene within a population signify a problem, which would trigger or activate natural selection to intervene? Which traits would natural selection target, and to what end?

Natural selection only helps those populations which are in trouble. But, what is the trouble with a net increase in the varieties of a particular gene? And if natural selection does become activated, how does that create new species? These are all logical questions that any normal inquisitor would ask, but science has no answer. I've found that whenever you trace any evolutionary claim to the bitter end and expect some sort of understanding, all reasoning vanishes.

———————

Occasionally, I will visit the question/answer websites to check out what kinds of questions the general public is asking about evolution, and what answers they are receiving. I ran across one that typifies the bulk of questions: "If evolution is true, why don't we see any in-between species, such as half-bird and half-lizard?"

Notice how this biologically savvy person who answered the question directs the petitioner's attention away from any kind of evidence of evolution, and opts for a more evasive answer.

"Most species that exist in competition with others for resources are all 'living examples of evolution,' since they're continuing to evolve.

If you're looking for a creature that's halfway between two existing classes or families—such as a bird that's "turning into" a lizard, as you suggest—then I don't think you have a clear understanding of what evolution is, or how it works.

Evolution is a change in genetics from one generation to another (mutation).

Those changes accumulate depending on their effect on survival and reproduction (natural selection), eventually resulting in divergent genetic lines that wind up as different species. As more changes accumulate, differences can widen, moving the species even farther apart in size, appearance, appendages, specific organs, etc."

Chapter 11

A Major Shift in the Theory of Evolution

The model that science has always held regarding evolution is that species evolved gradually over billions of years—a slow, progressive change from one species into the next in a line, as indicated in the graph below. This is known as gradualism.

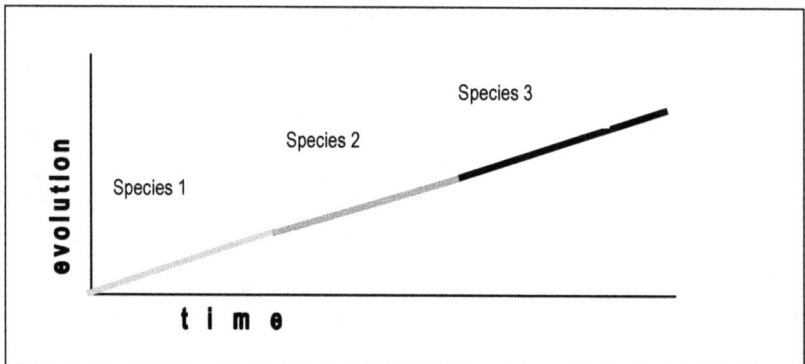

Gradualism

By the 1960s, the scientific community had accumulated enough prehistoric evidence from their digs to realize that species did not evolve gradually, as previously thought. Instead, they appeared to remain the same for millions of years, and then suddenly leap ahead to form a new species. But what could cause this strange phenomenon?

In the early 1970s, after 110 years of trying to get the discovery (the way things are) to fit the theory (the way they wanted things to be), paleontologists Steven J. Gould, Niles Eldredge and others began to rethink evolution, and design a theory that would actually correspond with these new findings.

Gould's scientific team figured that during periods of environmental stability, species remained basically unchanged. However, when threatened by food supply changes or extreme climate events, this would accelerate the rate of change in gene pool frequencies. Genetic drift with natural selection would cause periods of rapid evolutionary change, creating a radically different species. They also figured that mutations in regulator genes, which operate like master switches in orchestrating the development of body parts, could probably speed up this process. This became known as Punctuated Equilibrium.

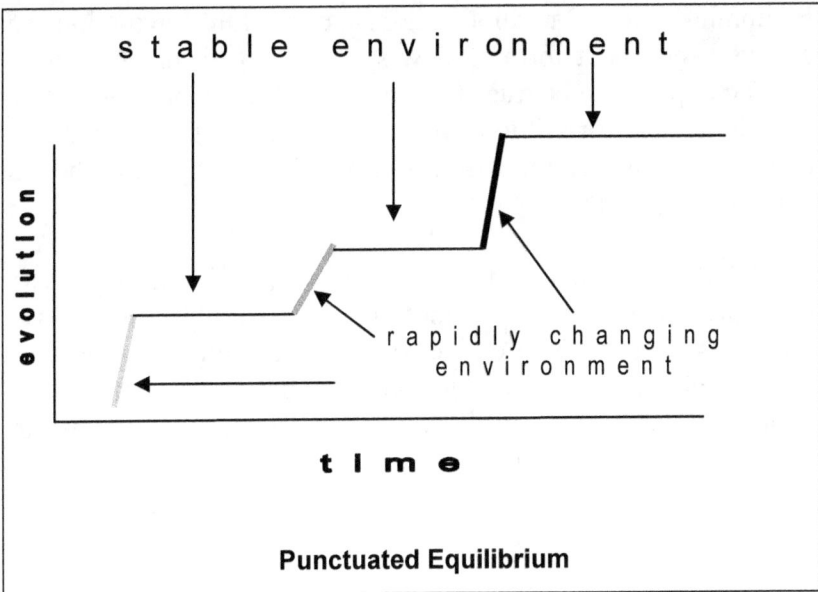

Punctuated Equilibrium

Debunking Gould's Theory of Punctuated Equilibrium

Paleontology's new theory may sound plausible on the surface—until you begin to ponder some of earth's more ancient residents, such as the elephant. This creature's appearance hasn't changed significantly since he came on the scene as a paleaomastodon, some 40 million years ago. Although natural selection has increased his size, and variations of him have appeared over the ages in the forms of stegodons, mastodons and woolly mammoths, they are still totally recognizable as the same animal.

Their trunks, tusks, large ears and "tree trunk" limbs have been consistent throughout their entire existence, and while they have lived through all of Earth's natural catastrophes and rapid climate changes, they have never evolved into a different species. This historic fact certainly contradicts what Dr. Gould's team theorized regarding the effects of climate change on evolution. In fact, the elephant's ancestry reveals an even bigger problem; one that speaks volumes against the entire idea of evolution. Beginning with Darwin's original conjecture, evolution has always been about forming new species, but history has shown just the opposite to be true. The numbers of different species have been in decline for millions of years, and in freefall for hundreds of years. We are continually losing species, not gaining them as evolution suggests.

In the elephant family, all but the African and Asian elephants have become extinct. In fact, twenty-four types of elephants have become extinct over the last forty million years and absolutely no new elephant species have emerged. If evolution were true, Earth would be teeming with new and different plant and animal life.

The latest in an ever changing opinion on how new species originated

Paleontology now believes that new species *(Newly discovered, not new species)* occur in two different patterns: Adaptive Radiation and Successive Speciation.

Adaptive Radiation *(meaning to spread out)* is where a species divide into two or more groups as they adapt to different environments. First, the group divides into distinct breeding groups. *(They can't really explain why this happens.)*

Next, because they're divided as separate groups, the gene pools of each group begin to acquire different mutations and random genetic drift. *(They can't really explain why two groups divided geographically would then begin to change in different ways, simply because they've been divided and now have different environments.)* And finally, because they've been separated, natural selection selects differently for each group until, over time, the two groups diverge enough that they can no longer reproduce with each other. At this point the two populations are each a different species. This branching pattern is known as Cladogenesis.

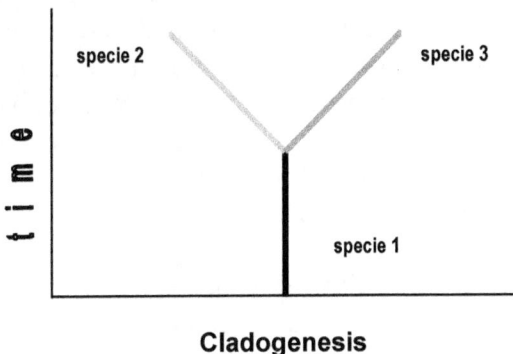

Cladogenesis

Debunking the Theory of Adaptive Radiation

This theory can easily be proven wrong using any number of examples, along with some ordinary common sense. But since we've used the elephant before, let's go with him again.

All elephants, even the extinct ones, can be traced back to a common ancestor — the Paleaomastodon. This early species *(about forty million years ago)* was about the size of today's pig but looked exactly like the elephant of today. Over millions of years, this animal has been geographically and socially isolated into many different groups *(gene pool populations)*. None of these isolated groups have ever evolved into anything but various types of elephants. In contrast to the "Adaptive Radiation" claims, any two elephants can mate and produce an offspring, and that offspring will always be an elephant.

Allow me to take this a step further in order to confirm my point. A team of scientists, in a joint venture between Japan and Russia, has been working together in order to re-create the woolly mammoth, which has been extinct for ten thousand years. During the recent global warming, areas of Russia have thawed, which have previously been permanently frozen; exposing well-preserved woolly mammoth remains. Live DNA has been obtained from these tissues. Now scientists will be able to implant woolly mammoth DNA into the egg of an African Elephant, which will act as surrogate, and give birth to the first woolly mammoth to exist in ten thousand years. The scientists claim this will be completed within the next three years. If this can be done, it should further invalidate the Adaptive Radiation Theory. Elephants can still "mate" with their ancient predecessors, even after millions of years.

Successive Speciation (spee-see-ā-shun) This is the second pattern of evolution that paleontology believes can occur. It is a single evolutionary line without the branching that adaptive radiation displays. In this pattern, instead of evolution being caused by species dividing into two separate environments, all members of a population remain as one. Some will react differently to their environment than others, causing differences in gene pool frequencies and mutations. Even though those members are still included in the population, they will become unable to breed with their prior kind, and will become a different species in time. This straight-line evolution pattern is known as Anogenesis.

Debunking the Theory of Successive Speciation

Successive Speciation Theory is actually identical to Adaptive Radiation Theory in the way that it works with one exception; under the successive speciation pattern the evolving species is not even isolated from the main population. This makes the claim even more preposterous than before. Regardless of how paleontology manipulates their theories, one thing remains constant — None of the missing links between any of the different species have ever been found. As long as this vital evidence remains missing, evolution will never be proven.

Chapter 12

Intelligent Design

The term "intelligent design" implies that a higher power has had a hand in the creation of all things in our universe. The mere utterance of these words to most scholars of the sciences can evoke a response similar to holding a crucifix before a vampire, while wearing a garland of garlic. And a few religious fanatics out there may even fantasize about driving a stake through the heart of any scientist who smugly dismisses the possibility of divine intervention.

In view of the ongoing conflict between the two factions, I do think the topic deserves serious examination. The fact is, there are a lot of questions that science simply can't answer. These are questions like: How did life begin? Any biologist, or fifth grader for that matter, can tell you that the first life on earth was a protozoan (single cell organism). But just how it got here in the first place is anybody's guess.

Once this single-cell organism came into being, it simply floated around in the waters of the earth for over a billion years before it mysteriously began to develop into other life forms. Or maybe it didn't develop at all. We do still have protozoa floating in the waters of the earth. Is it possible that God created a multi-cellular life form to feed on the protozoa? Could some divine power have been setting up Earth's food chain from the bottom up?

There are mysteries on this blue planet of ours that no scientist will ever solve—simple, common, everyday mysteries that baffle the bejeebers out of the most brilliant minds of our time. For instance, how does a monarch butterfly, emerging from its cocoon on a milkweed plant somewhere in Indiana, know that it is supposed to fly north for the summer and then south in the fall, in order to winter over in the Sierra Madre mountains in Mexico with every other monarch butterfly east of the Mississippi? And then in the spring, how does it know to fly north and plant its eggs on a milkweed plant somewhere in Indiana, like its parent did, before it dies in order to complete its life cycle?

Monarch butterflies don't fly in groups, and not one of them has ever made the trip before. This phenomenon is called migratory instinct, and it simply reeks of intelligent design. And here's another mystery: Even though no biologist can explain this phenomenon, why do so few admit to the belief in intelligent design?

There are easily enough examples of intelligent design to fill an entire library, but if one exceptional example is not enough to sway the mind of a doubter, then dozens would probably prove to be an exercise in futility. But in spite of that, let's look at another. Dr. James A. Shapiro of the University of Chicago discovered that a cell under stress will splice its own DNA into over 100,000 pieces. Then, a program within the cell senses hundreds of variables in its environment, and re-arranges those pieces to produce a new, more resistant cell. Could this be happening by pure chance, or is this more evidence of divine providence? Can one look around at all the miracles in everyday life and believe that random mutation, combined with sheer dumb luck, is the answer to life's mysteries?

One of the most convincing arguments for intelligent design came out of the mouth of a well-respected scientist in the field of biology. It came not as a statement supporting intelligent design, but in response to a question from an interviewer on a TV news program, in the form of a single word. That word was "no." But let me start at the beginning.

I'm a huge fan of the CBS TV news program *60 Minutes*. I watch it every time it's not being preempted by some sporting event. You can always count on a compelling interview with an interesting person that you just won't see in other news coverage. I had seen several teaser ads about an upcoming show featuring Dr. J. Craig Venter, world-famous microbiologist. All that week, I planned everything around seeing that program. Steve Kroft was going to be doing the interview, so I knew from past experience that he would be asking the same questions I would have asked.

Dr. Venter had just announced to the scientific community, in May of 2010, that he and his team had created the world's first synthetic bacteria with man-made DNA. This was indeed an earth-shattering accomplishment, one that the scientific establishment had been dreaming about for decades. The doctor had already distinguished himself as someone to watch in the world of science with the pioneering work he did in deciphering the human genetic code.

Kroft's interview revealed a man whose ambitions would not be denied. Venter had been booted out of his own organization after cracking the human genetic code a few years earlier, but he had made a comeback and trumped his old organization with this latest discovery. Now he had emerged victorious over his peers and didn't mind rubbing their noses in it. As Kroft pointed out, "He has made a career out of bucking the scientific establishment and earned lots of enemies along the way with his brash behavior and his knack for grabbing research money and the spotlight."

It became obvious in the interview that Dr. Venter was proud of his accomplishments, which included owning two research labs located on opposite coasts, where he employs a total of five hundred people. He also owns a 95-foot research yacht, as well as a home on the cliffs overlooking the Pacific in La Jolla, California.

Part of the interview took place speeding along in Venter's Aston-Martin. And while Steve observed that he was obviously an adrenaline junkie, his questioning also uncovered a scientist who undoubtedly loves life and is enjoying every day. When Venter travels the globe, collecting samples for his research projects, he always takes his wife, Heather, and their dog Darwin along. (Get the inference?)

The importance of Dr. Venter's discovery cannot be overstated. Being able to rewrite a cell's DNA will make it possible to bioengineer new medicines, foods and clean sources of energy, all in a world that is rapidly running out of options. Next year's flu vaccine could be created in a few hours instead of a few months. The possibilities are limitless.

Near the end of the interview, Steve asked Dr. Venter if he was playing God, which the scientist quickly denied. Then he asked him if he believed in God. Venter smiled coyly and said, "No." But I began to think about his responses to those two questions. In describing his discovery, Venter had said, "It's alive and self-replicating. That means it can indefinitely grow and make copies of itself." But all cells are alive and self-replicating. They all indefinitely grow and make copies of themselves. This is the definition of life. Dr. Venter can't take credit for a process that has been going on for two billion years.

What he can take credit for is replacing the DNA of a host cell with man-made DNA, so that when it does replicate, it reproduces the computer's generated DNA. In the view of the believer, when a man uses God's living cells to implant computer-generated DNA, and then denies the existence of

God—the one who provided the cells in the first place—that comes pretty close to playing God.

But the real irony is this: While Venter does not believe in intelligent design, it was only through the use of his hands and mind that he was able to "intelligently design" computer-generated DNA. His experiments have only proven that life did, indeed, have an intelligent designer.

———————

My grandfather died when I was eight years old. It was the first time I had seen a dead body, and it made an impression on me that I had difficulty dealing with. It was also the only time I had seen my father cry. In those days, when a man showed any emotion, it was viewed as a sign of weakness. My father was anything but weak, but because I had never seen him cry before, it served to heighten my own fear about the unknown.

Today, 65 years later, I can still replay the scene in my head...still smell the flowers that surrounded his casket...remember the glow of indirect lighting from the floor lamps on either end of his casket, softly illuminating features that resembled chiseled and polished stone more than human flesh. I suddenly realized the frailty of life for the first time and felt the fear of seeing my own father in a casket like that some day.

At the graveside service, the minister stood erect with an open Bible cupped in his hands. With a stoic expression, he quietly read the words I've heard so many times since. "From the earth we have come and to the earth we are returned, ashes to ashes, dust to dust."

I was confused. I knew my grandpa wasn't made of ashes, and I was sure he didn't come from the earth. But over the years, I have come to realize just how accurate those words really are.

Where do we come from, really?

About two and a half billion years ago, the earth began forming an atmosphere that would support life. It was comprised of three-quarters nitrogen and one-quarter oxygen, about what it is today, and formed a protective "bubble" or canopy around the earth, held in place by gravity. The atmosphere prevented anything, except for the largest of meteors, from hitting the earth. It also prevented water from leaving the earth. When water evaporates, it simply falls back to the earth as rain. So, the water you drink or bathe in today is actually billions of years old. There is no new water, only recycled water.

For the last two billion years, the earth has been an orbiting recycling center, and a greenhouse of sorts, traveling through the 460 below zero Fahrenheit temperatures of outer space at a speed of 67,000 miles per hour, in an orbit around its only source of energy: the sun.

But water is not the only thing that's two billion years old. Everything on the earth must be as old as the earth, since nothing has been allowed to enter or leave its atmosphere. Have you ever heard someone slam another with the insult "He's as old as dirt"? As it turns out...it's no insult.

The human body is comprised entirely of elements and minerals, the same ones that make up the earth's natural resources. More than 96 percent of our body weight is made up of oxygen, carbon, hydrogen, and nitrogen, with a nearly four percent combination of many other trace elements and minerals. Because we are a part of the earth, our bodies are as old as the material we're made of. Does this elaborate balancing act we call life sound unplanned and undesigned to you, or do you think that just maybe, someone might actually be at the controls?

Chapter 13

CREATIONISM

DNA: God's Fingerprints

George Washington, father of our country, is thought to have bred the first mules on the North American continent. It is said that he bred a large Andalusia, a male donkey named Royal Gift (which was a gift from the king of Spain), with his large brood stock of work mares. He eventually had fifty-eight mules working his Mt. Vernon farm.

What does this have to do with creationism? Because a horse and donkey are different species, every mule is born sterile. It has to do with one of God's laws. When God created the many life forms on this planet, He assigned each one of them a special numerical code. This code insures that God's commandment "after its own kind" will be honored forever.

Within the nucleus of every living cell are sets of chromosomes contained in the DNA that determines inherited traits. God has assigned a unique number of chromosomes to every living thing, both plant and animal. In order for anything to reproduce "its own kind," this numerical code between two breeding species must match, or the offspring will not be able to reproduce and the inherited traits cannot be passed on. This explains why every mule is born sterile.

Every living thing receives half of its chromosomes from its mother and half from its father in sets of two, so the number of chromosomes in each parent, when added together and divided by two, must result in an even number. If the number is not even, the offspring will not be able to reproduce. The mule is the only animal on earth with an odd number of chromosomes.

For example, a horse has 64 chromosomes, 32 sets. A donkey has 62, or 31 sets. When the two breed, $64 + 62 = 126 / 2 = 63$. This is the number of chromosomes a mule has. Since chromosomes are passed in sets of two, no matter what a mule is bred to, the result will always be an odd number—so the passing of traits stops with the first generation.

The Theory of Common Ancestry

The TV program Animal Planet has always been a favorite of mine, because I love animals so much. But when the crocodiles begin their bloodbath of a feeding frenzy, pulling helpless antelope under the river's tide to their death, the futile struggle to escape the crocs' gnashing, grinding teeth turns my stomach, and I have to turn it off.

But often, I catch a segment of Jane Goodall being interviewed with her chimps. And when I do, it seems she never fails to mention that we share 98 percent of our DNA with chimpanzees, an inference that we evolved from the chimp. In still another visit to the many question/answer websites, I read the following reply to someone's question about our relationship to the chimp.

"What we find is that chimpanzees share more than 98% of our genome; bonobos a little less and orangutans even less. The more the differences between the genome of any two animals on the planet, the farther away their common ancestor. We are more closely related to chimps because they are our closest living cousins. We share a common ancestor who lived about six million years ago. This is very impressive evidence for common ancestry (evolution) indeed."

Debunking the Theory of Common Ancestry

This thinking is typical for many science types. They think that it can easily trump any doubts one would have about evolution. The problem is, there isn't an ounce of truth in it. The fact is that the human genome is more closely related to a common mouse than it is to a chimp. In 2001, when scientists sequenced the genome of a mouse, they found their commonality to be a staggering 99 percent. Out of the 30,000 genes we share, all but 300 are identical. By using their calculations, this would mean that man is more closely related to the rodent, who lived 100 million years ago, than he is to that common primate ancestor they're always referring to, who lived only six million years ago.

This can only make sense from the view of creationism. DNA is God's fingerprint, and it's only natural that we would find them on everything He touched. Every living thing, both plant and animal, are made from the same elements and minerals. We all share genomic similarities because we were all created by the same God. Even the order of creation, found in the book of Genesis, has been confirmed by science. It is only common sense that every species, as created, would have to feed on an earlier species in order to survive.

Chapter 14

The Showdown

"A confrontation that forces an issue to a conclusion"

You have seen the evidence and claims from the perspectives of both evolution and creation. There comes a point when two opposing views must ultimately be judged in order to determine what is true, and what is not true.

Paleontology claims that humans and chimps share a common ancestor from six million years ago, which we will refer to as "primate" in the example below.

The exact point at which evolution is claimed to have occurred is called *divergence:* "To move in separate directions from a common point." This is the exact point when, according to paleontology, primate becomes man.

So, let's recreate and examine that point of divergence.

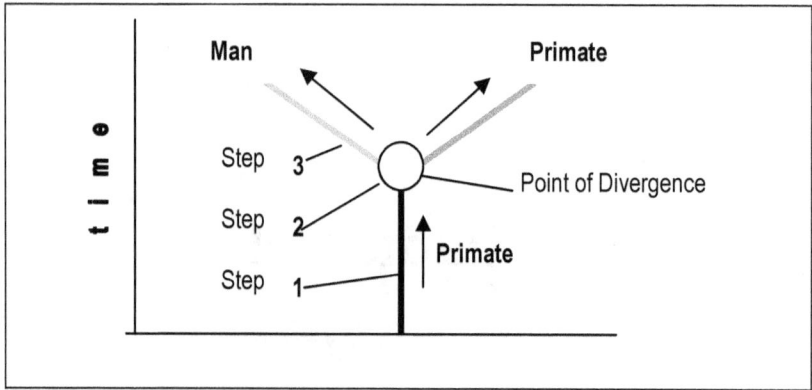

Step 1 – CLAIM: According to *Adaptive Radiation,* two groups of primates divide into two separate environments. (We can only assume this means geographical environments.)

Step 2 – CLAIM: Because the primates have divided as separate groups, the gene pools of each group begin to acquire different mutations and random genetic drift.

DENIAL: Being in separate geographic locations can't change mutations or genetic drift.

Step 3 – CLAIM: Because the primates have separated, natural selection chooses differently for each group, until over time, the two groups diverge enough that they can no longer produce with one another. A new species has evolved.

DENIAL: Only the number of chromosomes determines who can successfully breed. Remember the mule example? Only through breeding can the number of chromosomes change, and that is in the second generation. Divergence cannot change chromosome numbers. *A primate has 48 chromosomes.*

CLAIM: But suppose the chromosomes could change. If the primate's chromosomes increased by two (1 set), then 48 + 2 = 50 is the new number.

If this primate breeds with another from the group, then 50 + 48 = 98 is the new total. The chromosomes of the next generation would be 98 / 2 = 49chromosomes.

DENIAL: 49 chromosomes will be sterile. Must have an even number in order to pass on a complete set of chromosomes to the next generation.

CLAIM: But suppose we decrease the chromosomes. If the primate's chromosomes decreased by two (1 set), then 48 - 2 = 46 is the new number.

The primate breeds with another from the group, and 46 + 48 = 94 is the new total.

The chromosomes of the next generation: 94 / 2 = 47 chromosomes

DENIAL: 47 chromosomes will be sterile. Must have an even number in order to pass on a complete set of chromosomes to the next generation.

SUMMARY: As you can see, it would be impossible for two primates with a different number of chromosomes to produce an offspring of any kind, let alone a human.

The Theory of Chromosome Fusion

Realizing the chromosomal disparity that existed in their theory on how new species evolved, scientific minds contrived still another maneuver.

(Science does not know whom the ancestor from six million years ago was, which was common to both man and chimp, so we are using the chimp's chromosome set-up to illustrate this theory.)

If you have never seen a diagram of a chromosome, imagine an elongated tube with bands around it. The bands represent the alleles that determine the character traits.

In this latest development, science noticed a strange similarity between the human chromosome #2 and two of the chimp's chromosomes. If you trimmed a band or two off the bottom end of one of the chimp's chromosomes and off the top of another chromosome, and then fused them together, the new chromosome not only looked exactly like the human chromosome #2, it also reduced the chimp's chromosome count to match that of a human—46 instead of 48, or 23 pairs. They reasoned that somehow, through a mutation, the two chimp chromosomes had become fused together as one to create the first human.

Debunking the Theory of Chromosome Fusion

This latest theory manipulation is fraught with problems. First, a few of the chimp's gene bands would have to be discarded in order to become a perfect match. But the bigger problem is this: Assuming this complex fusion could be the result of some mutation, this altered chimp would still have to locate a chimp of the opposite sex who had experienced the same exact mutation, involving the same chromosomes, in order to mate and produce the earth's first human. And then, who would the offspring of this union mate with in order to establish this new family of human species? If everything happened the way science proposes, it would be a clear indication that divine intervention was involved.

Mankind's earliest ancestors?

Even before Darwin's publishing of *On the Origin of Species* in 1859, curious supporters of his theory began exploring the earth, hoping to find the evidence linking man to his suspected ancestor, the ape. Every shovel of dirt that was turned brought great expectations of finding man's origins.

In 1856, quarry workers had found fragments of what would become the first Neanderthal ever found, named for the find location, Germany's Neander Valley. The remains were ultimately turned over to a local naturalist, Johann Carl Fuhlrott, who realized that they were possibly early human remains. This fossil material was eventually dated to be about 40,000 years old.

Since that first find, digs the world over have revealed thousands of pieces of evidence, mostly of the primate order, covering nearly seven million years. These finds have all been categorized into various hominid types, or more correctly hominidea (haw-MIN-uh-dee), a family of the mammalian order of primates. Paleontology's data on hominids is the most reliable source of information today.

Understanding the Chronology of Human Development

The purpose of the chart below is to allow you to see how man developed, not based on the theory of evolution, but based entirely on fact.

Primates appeared suddenly 65 million years ago and have continuously survived until today, as the Hominids line indicates.

Although paleontology classifies early man (homo sapien) among the 23 known hominid sub-species, which have lived at various times over the last seven million years, early man is *not* a hominid, but the first known human. But with paleontology, there is no distinction made between primates and humans, since they believe humans evolved from primates. They are both simply considered part of the mammalian family, designated with the letter "H," or Homo. The word sapien means "wise." So, homo sapien means "wise hominid" or "knowing man." But the chart reflects the fact that man appeared suddenly two million years ago, not seven million years ago as a hominid.

Although God created both humans and primates, the distinction between the two should be obvious to the most casual observer among us. Man thinks, speaks, and attempts to conquer everything in sight. The primates' only concern is survival. Man also used stone tools, where the primate did not. But the most important difference is this: Primates have 48 chromosomes and humans have 46. This means that no matter how incessantly science tries to link the two together as one through evolution, primates and humans have never produced an offspring. It simply isn't possible, as we have seen. So using this criteria, our line for early man begins two million years ago and continues until they became extinct, some 150,000 to 200,000 years ago.

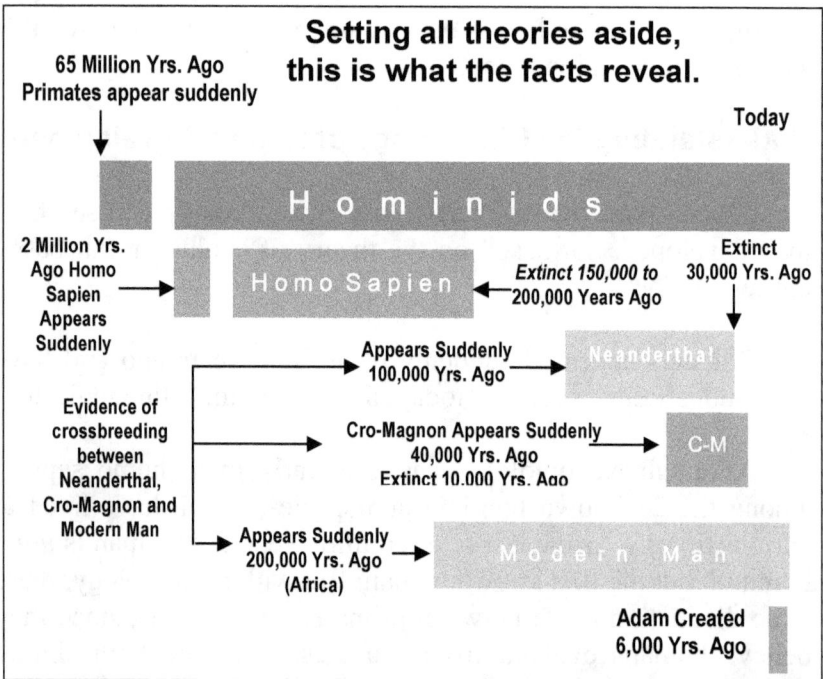

Setting all theories aside, this is what the facts reveal.

65 Million Yrs. Ago Primates appear suddenly

Today

H o m i n i d s

2 Million Yrs. Ago Homo Sapien Appears Suddenly

Homo Sapien

Extinct 150,000 to 200,000 Years Ago

Extinct 30,000 Yrs. Ago

Appears Suddenly 100,000 Yrs. Ago

Neanderthal

Evidence of crossbreeding between Neanderthal, Cro-Magnon and Modern Man

Cro-Magnon Appears Suddenly 40,000 Yrs. Ago Extinct 10,000 Yrs. Ago

C-M

Appears Suddenly 200,000 Yrs. Ago (Africa)

Modern Man

Adam Created 6,000 Yrs. Ago

About 200,000 years ago, before the last of the 23 homo sapien types became extinct, Anatomically modern man, known as homo sapien sapien (very wise hominid) appeared suddenly on the African continent, indicated by the line marked "Modern Man". Whether this newcomer had anything to do with the extinction of homo sapien is unknown.

Then about 60,000 years ago, modern man migrated into Europe and settled around the Mediterranean. Evidence shows they began integrating and interbreeding with Neanderthal populations who preceded their arrival by about 40,000 years. Modern man's better agility, longer limbs and superior hunting skills probably had more to do with the extinction of the Neanderthal people than any other factor.

You may recall seeing pictures of Neanderthal man, the original cave man, in your school science books. So you may be wondering, "Why would God have created such a crude individual to represent mankind?" The answer is simple. God created Neanderthal exactly the way he needed to be in order to survive the cruel conditions of Earth during his time. His deep, thick chest cavity served to better insulate the body's vital organs from the cold in the northern latitudes during the earth's glacial phases. His superior upper body strength was needed to throw heavy stone spears to kill his prey. His short legs and squatty stature were adequate, since he could hunt from under the cover of heavy forestation and didn't need to run down his prey. There is no way that the much frailer body style of modern man could have survived under the more hostile environmental conditions of their time.

In the meantime, God created Cro-Magnon, who coexisted with Neanderthal during the last 24,000 years of his time on Earth. There is evidence to support the fact that there was interbreeding between these two groups of ancient humans. A small amount of Neanderthal's DNA has been found to be present in the genome in most populations of today's humans. But never has there been any DNA from the Hominids found. This further confirms the fact that mankind did not evolve from apes. The natural propensity for all humans to mate with whomever is available is ingrained and pervasive. The crossbreeding activity accounts for the introduction of all the modern world's races and ethnicities. Check out the chart above for more detail.

While there has been more attention given to Neanderthal than to Cro-Magnon, neither genus seems to have a clear predecessor, which certainly smacks of creationism to me. Both of these homo sapiens were more advanced than previously given credit for. Cro-Magnon's body style was closer in appearance to the anatomically modern human, with the exception of Cro-Magnon's larger body frame and cranium. While both were cave dwellers, the caves in France, Spain and Australia where Cro-Magnons were found displayed artwork that would give today's artists a run for their money.

Summary

In this chapter, I have taken you right to the heart of the final two versions of evolutional theory, Adaptive Radiation and Chromosome Fusion, at that moment in time when one species transforms into another, and proven them both to be wrong based on biology's own genetic, reproductive principles. Yet the teaching of these flawed theories continue in our public schools.

When a theory continues to be taught to young minds beyond the point that its authenticity can be defended, it moves from being credible curriculum into being planned deception, whether realized or not.

After five generations, the teachings of evolution are so indelibly fused into our nation's science curriculum that it continues to flourish like a tumor, choking off all logic and sound reasoning, passing off its flawed precepts as gospel to still another generation. However, in defense of the scientists who wrote today's teaching materials, they only know what they were taught—and other scientists who only know what they were taught, taught them. Besides, most scientists are afraid to question authority for fear of losing funding and being ostracized by the scientific establishment.

A Refreshing Exception

In spite of the broad acceptance of evolution among members of the scientific community, I have been able to locate a few well-credentialed and respected scientists who believe strongly in creationism. One in particular, Dr. Ed Neeland, has been gracious enough to allow us to reprint his opinion here:

Dr. Ed Neeland is a published scientist and chemistry professor at the University of British Columbia with Honours in Bachelors of science. He has a double major in biology and chemistry. He has a Masters of science in organic chemistry and a doctorate in organic chemistry, with a post-doctoral in botany.

He writes the following:

"I have a good reason as a professional chemist, to question evolution's version of origins. On occasion, several of my untenured colleagues have slipped into my office and said they too didn't buy into it, but were not about to make their opinions public. Fair enough. So why do I hold a diametrically opposed position about evolution?

My area of research is synthetic, organic chemistry. What this means is that I make new compounds. They could be new medicines, plastics or explosives, to name a few. Please understand that very exacting conditions of concentration, time, temperature, atmosphere and purification are needed to make these molecules. I have to work very hard to make even moderately sized compounds, and the "simple" cell is replete with complicated compounds that, frankly, I couldn't begin to synthesize. And yet, evolution demands that life began using unguided, uncontrolled chemical reactions under unexacting conditions. I don't see it, not at all. I am trying hard not to use the term nonsense. I have failed.

Thousands of intelligently designed and carefully controlled experiments repeatedly cannot replicate the components of the cell nor explain, once made, how or why they would self-assemble. To continue to believe that random uncontrolled chemical reactions somehow made a living cell is to step into the world of a badly supported faith; maybe even delusion. But let's be gracious and grant that a living cell did happen somehow through natural events. What process enabled the cell to ultimately begin building up information to make new complex organisms? Apparently mutations. Assuming this process happened by itself, then surely intelligent scientists should be able to take a simple bacterium and through mutations, create a multicellular organism like a worm or an ant. Our efforts are not even close.

Something simpler? Take an organism and through mutations create a new macro-part. For example, a fruit fly with a stinger. We have been subjecting fruit flies to over a hundred years of mutagenesis experiments. Result? Fruit flies with different eye colours, wing styles, body types and even some fruit flies with legs growing out of their eyes. This is not generating new information or new parts; just scrambling the information that was already there, and it's not evolution.

I am forced to draw a conclusion. Repeated experiments disprove the very foundation of evolution, so why trust its claims? I do not. You could just as well claim that an iPod happened by itself through a series of lucky combinations of oil deposits, minerals and lightning, and that you will work out the details later. Good luck with that.

On the plus side, all that complex information came from somewhere, and again our experience shows that complex information always comes from intelligence. No exceptions I can think of. That well of complex information is the mind of God.

The Big Business of Evolution

Evolution is a multifaceted, multi-billion dollar industry, promoting, marketing, distributing and teaching something that has never been proven to be true. Its tentacles, through the use of multimedia, reach every man, woman and child with its agenda of indoctrinating and educating generation after generation of Americans in order to maintain its status quo. This requires a strategy of behemoth funding methods, which it excels at. Their grant writing operation, which taps into both taxpayer dollars and private wealth, is a major enterprise in its own right.

The National Science Foundation, where the business of evolution receives most of its funding, is funded entirely by taxpayer dollars and has a 2013 budget of $7.373 billion, an increase of 4.8 percent over the 2012 budget.

An independent industry, which has sprung up in order to feed off the "barnacles" of the host, is a system to educate the educators. You see, it takes a special system in order to market the incomprehensible, and do it in an orderly fashion so that it appears to be understandable. If you were to search the Internet for "Evolution teaching resources", you would find no less than 670 listings of organizations that produce and market selling aids for use in our classrooms nationwide. Here is a typical listing:

"Understanding evolution for teachers."

This site was created by the
University of California Museum of Paleontology
with support provided by the
National Science Foundation (grant no. 0096613)
and the Howard Hughes Institute (grant no. 51003439)

Notice that even their website is being paid for by grant money.

"Pay no attention to that man behind the curtain"

Exactly who is it that continues to sell our nation's children on the virtues of a false theory about man's origins? Who is it that actually pulls the strings that determines what will be taught, and what will not be taught? I only wish that I could answer that question, but I think it would take another one of those infamous federal investigations prompted by an act of Congress. Don't hold your breath on that one. You may have to settle for what I have found while doing the research on this book.

You would think that someone with the complete authority to mold our children's minds in any fashion they see fit would be someone selected by a steering committee and appointed by the President of the United States, and further subjected to a system of checks and balances with complete transparency...but you would be wrong. Actually, the system in place to keep our children in line and thinking the way they want them to think is very obscure and low-key.

It all began when the Russians launched Sputnik, the first unmanned satellite into space, in the summer of 1957. The U.S. government became concerned that we were falling behind in science. That concern initiated the funding of the Biological Sciences Study by the National Science Foundation. Even though evolution has absolutely nothing to do with space exploration, it soon became the main thrust behind this group of biologists.

Headed up by herpetologist Arnold Grobman and geneticist Hiram Bentley Glass, their first steering committee in 1959 decided to target high school biology, mainly the tenth-grade level, and brought out textbooks in 1961 emphasizing evolution as the major scientific theory. They have remained committed to the teaching of evolution since the study's inception.

Today, this organization, now known as the Biological Services Curriculum Study, or B.S.C.S., which became an independent non-profit organization in 1973, is more committed than ever to see that the evolution "illusion" continues to be strongly administered to prevent it from unraveling. As long as the minds of all future generations are implanted with the belief that we evolved from apes, the evolution organization can continue to flourish financially, sucking major funding from international sources that could be going to worthwhile projects.

Epilog

In the subtitle of this work, I asked the question: "Does scientific evidence support the existence of a Divine Creator?" The answer is a resounding YES!

There is an old coin toss trick that will ensure a win for the tosser every time. You simply say, "Heads I win, tails you lose." When we were kids, my older brother used to win tosses regularly, until I finally woke up to what was happening. He was winning *directly* when he tossed heads, and winning *inversely* when he tossed tails. The same principle is true when presenting evidence to support the existence of a Divine Creator, but without the deception.

I presented *direct* evidence to support the premise in Chapter 3, when I pointed out that geologists had found evidence of great flooding in the earth's strata, covering a 2000-year period ending just before the fourth millennium B.C. This corresponds exactly with the flooded condition of the earth as described in Genesis 1:2, just prior to the creation of Adam. This was just one of many examples of *direct* evidence given throughout the book.

But I also presented many instances of *inverse* evidence, such as in Genesis 1:1, which states, "In the beginning, God created the Heaven and the earth." Since this Biblical statement does not indicate the time of this occurrence, it follows that it could correspond with science's determination that the universe is billions of years old.

Another example of *inverse* evidence is science's dating of the various species on Earth and their order of appearance. While science claims this was due to evolution, it also matches the order in which God claims to have created them.

See Genesis 1:11 through 1:27.

But there is not one single example that any person can name in which factual, scientific evidence disputes what is stated in the Bible when properly interpreted. This leads me to only one possible conclusion: God, in His infinite wisdom, has created a pretty wonderful place for a whole lot of (fill in the word) people. When evolution is the belief of choice in a world of everyday miracles, God help us.

Post script: Looking in retrospect from where we are today, it's easy to see how we ended up having creationism denied as an acceptable curriculum. First of all, creationism is viewed as a religion. But creationism is not a religion, any more than evolution is a religion. Religion is a man-made institution whose only purpose is to praise God. Creationism is a theory, just as evolution is a theory, dealing only with how the universe and Earth's inhabitants came into being.

Unfortunately, the only ones interested in promoting creationism are the various sects of organized religion. That opened the door for the Supreme Court justices to declare it in violation of separation of church and state, as they understood it to be defined in the U.S. Constitution. It is my belief that, if a purely secular group were to file suit against the U.S. Department of Education and fight the battle all the way to the Supreme Court, petitioning them to consider creationism as suitable curriculum for our public schools, as long as religious beliefs were eliminated, I don't see the justices as able to ban creationism from being taught in our public schools. But, of course, I'm the eternal optimist.

About the Author

Let's begin this bio by giving you a quiz.

1. Who was Pavarotti's voice coach?

2. Who taught Mark Spitz how to swim?

3. Who was Barbara Streisand's third-grade music teacher?

You don't know who mentored these famous people?

My point is this: In every one of these champions' lives, there was someone who truly made a difference. Without their help, encouragement and influence, these people would very likely have failed in life. They may have turned to alcohol or drugs, or just given up.

But you see, like these celebrities' mentors, who I am isn't important. What I know is.

I've been preparing to write this book my entire life. I was raised in a loving home but struggled with school. I've worked for a living, and I've known despair as well as hope. I've been defeated, but I also enjoyed some victories along the way. I've experienced happiness and sadness. I've loved, lived, and raised a family. I lost my only love to cancer ten years ago.

These are my life experiences. They define who I am. But they carry no credentials that I can affix to the end of my name in order to impress you. I am not an author, a scientist or a clergyman. I'm only me—but I could be a lot like you.

I have never asked God why, or why not, because He owes me nothing and I owe Him everything. He sent his angels, agents, spirits, guides, or whatever you'd like to call them, to help me write this book. They guided me in my research and answered questions before I asked them. He neglected to tell me how to write my own bio—but I think I know why.

You are the celebrity in your *own* life.

And this book isn't about me. It's about what you need to know. . . the truth.

May each of your life's endeavors be an "awakening."

C. Robert "Bob" Follett

Special Supplement
Part I

The AWAKENING
"Does Scientific Evidence Support the Existence of a Divine Creator?"

Unlocking the Hidden Truths of Genesis

Are you a freethinker? A freethinker is one who forms opinions based on science, logic and reason—not influenced by authority, tradition or any dogma.

As a freethinker, I believe truth always trumps tradition. So, when I decided to research the greatest human divide, that of science versus religion, I promised myself that I would accept whatever I found as long as it agreed with factual evidence.

I was simply amazed to find that scientific evidence actually supports the Theory of Creation, *not* the Theory of Evolution, as one would imagine. But even more surprising is the fact that this same evidence does *not* agree with religion's interpretation of the Biblical account of creation, as well as other important teachings in the book of Genesis. While religion's doctrine has prevailed for nearly two thousand years, it can only be true if it matches the evidence—the same evidence that science ignored when promoting their Theory of Evolution.

I have found that many theistic (believing) people are more than willing, even excited, to learn that my book soundly disproves the Theory of Evolution, using scientific evidence and common logic. But when I turn the microscope on the Holy Bible, the impenetrable defense shield emerges to protect beliefs they have blindly accepted through faith their entire lives. My purpose is not to discredit religion, but to strengthen its truths by using this same scientific evidence to prove that when the Bible is properly interpreted, science and religion are actually one.

Finding truth

Clearly, Genesis is the most questioned book in the entire Bible. This is because the stories of Genesis, which are being told from the pulpit of every church, synagogue and temple, do not agree with what scientific evidence has shown to be true. The earth's age is not 6,000 years old. God did not create the universe in six days, and Adam was not the first human on earth.

Because this 2,000 year old, literal interpretation doesn't correspond with factual evidence, it makes the Bible appear to be untrue. Consequently, many of today's most brilliant scientists have completely disavowed the same God that they prayed to on Sunday mornings when they were children. After all, how can creationism be true if the Bible is not true?

Religion's refusal to break from tradition simply widened the gap of human divide, allowing more people to fall headlong into the foaming brine of disbelief.

Five simple keys that unlock the true meaning of Genesis

As I was gathering research material and praying regularly for guidance to find the clues that I knew must exist, in order for the Bible to be true, information was revealed to me in a way that

I can only describe as being divinely given. —*I truly believe that each of us have spiritual guides who possess great wisdom beyond our understanding, and help us navigate our physical lives on earth.*

My spirit guide doesn't give me the answer, but instead, implants a question that serves as a clue, in order to guide me to the answer. The questioning doesn't stop until I finally realize what I am intended to know. When I get it right, the questioning stops. I don't literally hear a voice. It's more like hearing my own thoughts, but with a strange insistence similar to having a song stuck in your head.

The question I was given was... "Who wrote the book of Genesis?" I can remember thinking..."Why would I ask myself a question for which I already knew the answer?" When I gave the obvious response, "Moses wrote the book of Genesis." Spirit simply asked the question again..."Who wrote the book of Genesis?" You will see in the first "key" below, the answer that finally quieted my spiritual inquisitor.

I am listing these revelations, or "keys", in the order that I received them. Unfortunately, I wasn't given the fifth key until well after my book was published, thus the necessity of writing this supplement. I consider key #5 to be the most telling key of all; since it links Genesis to the New Testament and reveals what Christ's crucifixion was really all about.

Key #1 – Realize that God, not Moses, wrote the book of Genesis.

In the sixty-six books, which comprise the Holy Bible, the period from Adam's creation to Christ's crucifixion, represents 4,031 years. Of that total period of time, the book of Genesis makes up 60%; the other sixty-five books combined cover only 40%.

While Moses has been given credit for authoring the first five books of the Bible, his birth did not occur until the beginning of the second book, Exodus. This can only mean that, when Moses wrote the book of Genesis, he had to depend entirely on what God told him to write, since he wasn't even alive to witness what he was writing about. This clearly proves that, when we read Genesis, we are actually hearing the word of God, not the word of Moses. Moses was merely transcribing for God. This is not true of any other book in the Bible.

The chart below will clearly illustrate that Genesis is God's foundation for the other sixty-five books of the Bible. It only stands to reason that any misinterpretation in the book of Genesis would have an effect on the rest of the Bible, just as flaws in the foundation of any structure, whether actual or metaphorical, must affect anything that rests upon that foundation.

Bible Timeline

Adam's Creation 4004 BC			Christ's Crucifixion 27 AD
	The word of God **1 Book – 2434 years**	**The word of man inspired by God** *65 Books – 1493 years*	
	Genesis	Exodus - Revelation	
	60% Could misinterpretation here . . .	**40 %** make a difference here?	

Moses' birth

Key #2 – Acknowledge the missing time between verses 1:1 and 1:2

One of the most grievous ways Genesis is being misinterpreted is by refusing to acknowledge the obvious missing time, which exists between the first two verses.

Genesis 1:1 – *In the beginning, God created the heaven and the earth.*

Genesis 1:2 – *And the earth was without form and void. And darkness was upon the face of the deep. And the spirit of God moved upon the face of the waters.*

We know with complete certainty that the universe was created 14.5 billion years ago. If this is not when God is referring to in 1:1, then the Bible would have to be wrong. But since God isn't giving us a specific time for creation, we must assume that "the beginning" refers to 14.5 billion years ago. However, in verse 2, God is describing the condition of the earth just before creating Adam. We know that Adam was created in the year 4004 BC. *(See the Awakening, pages 33-39. Also Chapter 4, page 22.)*

There are numerous ways to prove that there are billions of years missing between the two verses, but I will list only three.

1. We know that Satan could not have existed before creation. And we also know that God left him in charge of the earth sometime prior to Adam's creation. He did, after all, appear as a serpent to tempt Eve in the Garden. Therefore, he had to exist somewhere between verses 1 and 2.

2. Science has confirmed that there were dinosaurs that roamed the earth for 150 million years. This had to be following creation in verse 1, but millions of years before Adam's creation.

3. Science has proven that modern man (gentile) first appeared from 200,000 to 250,000 years ago on the African continent and began migrating all over the earth 60,000 years ago. DNA from these early humans has been found in our genome makeup today. Visit www.ramsdale.org/dna10.htm for proof of this fact. This is still more evidence that a huge gap of missing time

exists between verses 1 and 2. The most conservative scientific estimates claim the earth's human population was between 5 and 7 million when Adam was created in 4004 BC. Some estimates range as high as 86 million.

Visit: www.tentmaker.org

It is quite apparent that God did not see fit to inform Moses of these early gentile predecessors. As we know, gentile is the biblical term for all nationalities other than Hebrew. This fact also answers the question: Where did Cain find his wife? She was a gentile.

When we ignore this expanse of time between the two verses, earth becomes 6,000 years old, and Adam becomes the first human. This is the mistaken interpretation that all monotheistic religions still adhere to today.

Key #3 – Change the word "man" to the word "Hebrew"

Here is an amazing fact that has been completely overlooked for 2,000 years: The word "man" appears 95 times in the book of Genesis. Each time, it is either used to refer to Adam or persons in his lineage. Since it is clear that Adam and all his descendants were Hebrew, *(This is proven beginning on page 57 of the Awakening, under the subheadings "Old testament teachings" and " Unveiling the truth.)* then the word "man" must be understood to mean Hebrew, not mankind. This confirms the fact that Adam was the first Hebrew created and not the first human.

It must be understood and remembered that the book of Genesis is a history of the Hebrew nation from Adam's creation, to the bondage of his descendants in Egypt. It has nothing to do with the history of the gentiles.

Key #4 – Consider who Moses was

In order to understand God's choice of words in Genesis, we must first understand who Moses was. If it were possible to place Moses in a time machine and bring him into the 21st century, we would soon realize that his knowledge about the workings of the universe would be no greater than that of today's six-year-old child. God would have needed to talk to him, using only the vernacular he would understand.

So, how would you explain the creation of the universe to a six year old? Probably something like:

"And God said, let there be light: and there was light. And God saw the light, that it was good: and God divided the light from the darkness. And God called the light Day, and the darkness he called night. And the evening and the morning were the first day."

There was no way God could have said:

"Moses, right now you're standing in total darkness. But, twelve hours from now, the earth will have rotated by 180 degrees, and you'll be standing in the daylight. This is because the part of the earth you're standing on now will no longer be in darkness, but facing the sun."

God didn't need to explain how he divided the daylight from the darkness, only that He was the one responsible for doing it. Moses also could not have comprehended the time represented by the phrase, "fourteen and a half billion years ago." So God explained time in terms that Moses could understand. Moses knew that a day was from sunrise to sunrise, which would include "the evening and the morning."

So, we pick up the Bible in the 21st century and try to read literally, what God was saying metaphorically. Remember—God wasn't talking to a physicist. He was talking to Moses.

Key #5 – Understand the meaning of "Image and likeness of God"

It would only be natural for people today, to believe that we were made in God's image and likeness, but how overtly presumptive it would be, to believe that God was talking about us.

Genesis 1:26 – *And God said, let us make man (Hebrew) in our image and likeness: ...*

Since we know God was not talking about creating mankind (gentile) in his image and likeness, whom He had already created 200,000 years earlier on the African continent, it should be clear that He was speaking of creating Adam after His image and likeness.

So, we have to ask ourselves—what is God's image and likeness?

First, we know that God has a physical body. He walked, talked and ate regular food with Abraham and Sarah. This is evident in Genesis 18:1-33. *(See the Awakening: Did you know... on page 20)*. As we know in scripture, Adam (the original Hebrew) was given an extended physical life, which the gentiles never did have. *(See the Awakening, "An ageless people" found on page 63.)*

God also has a soul. The soul is the seat of consciousness, credited with the faculties of thought, action and emotion. In us, it defines who we are as an earthly person. *(Webster's definition)*

But, God also has a spirit. Jesus made it perfectly clear many times that the spirit enables resurrection. It is impossible to inherit the kingdom of heaven without it. It is clear that this is the one thing that the gentiles did not possess, so there was no afterlife for mankind previous to the creation of Adam. *(Webster–*

Spirit: That portion of the human psyche that prevails at the time of death.)

This belief can be biblically supported in two ways:

First—God said that the Hebrew would lose both his spirit and his longevity of life after the flood. This statement is key.

Genesis-6:3 *And the Lord said, My spirit shall not always strive with man, for that he also is flesh: yet his days shall be an hundred and twenty years.* (The latter part of this statement referred to Moses' lifespan, which would be exactly 120 years.)

Obviously, if the Hebrew lost his spirit *after* the flood, then he would have had a spirit *before* the flood. This proves that the original Hebrews (before the flood) went directly to heaven upon their physical death. This is confirmed in:

1 Corinthians 15:45 *And so it is written, The first man Adam was made a living soul; the last Adam was made a quickening spirit.*

Second—This also suggests that the Hebrews born after the flood did not go to heaven, but rather "slept" awaiting a Savior.

1 Corinthians 15:20 *But now is Christ risen from the dead, and become the first fruits of them that slept.*

So, what caused the Hebrew's loss of spirit and physical longevity? The only thing different after the flood than before the flood was that, all the other Hebrew were dead, and the survivors on the Ark had to marry and have children with gentiles, who had neither a spirit nor longevity of life.

This also confirms that the flood could not have been a global flood as we have always been taught to believe, but a regional flood, designed to kill only the Hebrews. If only Hebrews survived a *global* flood, we would all be Hebrews today. Of course, this is not true. *(This is proven in the Awakening on page 60 "Not a global flood.")*

This would also explain the diminishing longevity of the Hebrews following the flood, since they would be breeding with the gentiles, who had life expectancies ranging from thirty to forty years at that time in history.

Clearly then, the purpose of the Savior was to reinstate the Hebrews' spirit by retroactively giving the gentiles a spirit, thereby providing a pathway to the afterlife for all mankind.

The ~~good~~ better news of Jesus Christ

We have all had the front door visit from a couple of well intentioned missionaries, asking us to repent and follow the way of the Lord, in order to avoid the implied certainty of death and destruction in the fiery furnaces of hell. Other speakers of persuasion take a gentler approach, by giving you a list of requirements one must meet, in order to be "saved" or "born again". But in light of what we have just learned, by re-interpreting Genesis using a more reasoned approach, it would seem that salvation has been unconditionally assured through Christ's crucifixion. If we are indeed, saved by the grace of God, as Paul has preached throughout the New Testament, how then can man hold God's grace hostage, releasing it only when certain conditions have been met? It would seem to me, that God has provided universal salvation as a way to bring His "chosen people" (Hebrew) home.

More on this topic at:
http://tentmaker.org/site-search.html?q=Universal%20salvation

Universal Salvation - I realize that universal salvation is a concept, foreign to most Christians, if not unthinkable. But let's consider it for a moment.

When the Hebrew lost his spirit after the flood, he went on functioning as before, which proves the spirit is not the soul. The soul is the seat of consciousness and facilitates thought, action and emotion; it is the soul that controls what the body does. So, our soul determines whether we will do good deeds and live righteously or commit sin. But the soul cannot be saved or born again since it dies with the body.

Since we know the Hebrew sinned before he lost his spirit—Adam and Cain are both proof of this—and continued to sin during the time he had no spirit, *(from after the flood to Christ's crucifixion)* then obviously, the soul is committing the sin, not the spirit.

On the other hand, the spirit is the only one of the three that survives death. The spirit is a gift from God, made Holy by Christ's crucifixion, and is therefore incorruptible. It has been sanctified by God and is incapable of committing sin. Does it make sense to punish the spirit for sins committed by the soul?

We are all saved by the grace of God, not by the deeds of man. I think Paul made that perfectly clear in his assessment below.

Galatians 2:21 – *I would not frustrate the grace of God: for if righteousness come by the law, then Christ is dead in vein.*

Long ago,
in a distant galaxy
far, far away . . .

I'll leave you now, with an interesting concept. If heaven is a physical place, perhaps a planet located near the origination point of our universe, 14.5 billion light years away—how could we ever send an exploratory ship to examine it?

Physicists have dreamed for years, of being able to travel at the speed of light. (186,000 miles per second) But even at this speed, it would take 25 thousand years, just to reach the edge of our own galaxy. There are billions of other galaxies in the universe. Its vastness boggles the mind.

As light travels, it is not immediate because it has mass. In other words, light must push the protons out of the way as it travels, which greatly reduces its speed. However, the human spirit has no mass. So theoretically, it should be able to travel anywhere in the universe and arrive immediately. Could this explain the near-death experience (NDE) phenomenon, where survivors describe going to heaven and returning during their "death", lasting only a few minutes, and without any brain activity recorded on an E.E.G.? It is claimed that as many as fifteen million Americans have experienced NDE; many were un-saved individuals who didn't even believe in God before their unexpected trip.
Visit: www.nderf.org/

One similar case is of a young boy named, Colton Burpo, who died at age four when his appendix burst. He later told his mom and dad, that during his NDE, he met his dead sister in heaven. Problem is, Colton never had a sister, at least not that he knew of. His parents had never told him that his mother had miscarried a little girl a year before Colton was born.

Colton, now age eleven at this writing, has written a book about his experience. It's called "Heaven is real" and is available at any book source. This incredible story has now been made into a movie.

So, my question to you is this:

If Colton's story is true, doesn't it give credible sustenance to a belief in universal salvation?

Special Supplement
Part II

The AWAKENING
"Does Scientific Evidence Support the Existence of a Divine Creator?"

The Flawed Theory of Evolution

FACT: Although science has developed a vernacular, which include words like *gradualism, punctuated equilibrium, cladogenesis, adaptive radiation and successive speciation,* — terms used to suggest that the human species came on the scene in degrees or incremental steps — scientific evidence clearly shows that the oldest human found was 100% human; not part human and part something else.

INDISPUTABLE CONCLUSION:

The first human was . . .

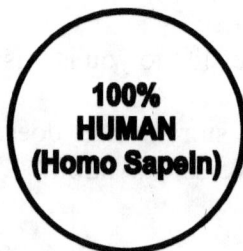

100%
HUMAN
(Homo Sapeln)

144

FACT: Assuming the first human was not created, then the first human had to be born, since there is no other way it could have arrived.

INDISPUTABLE CONCLUSION: **The first human was born.**

FACT: Since the first human was born a 100% human, then it had to be either male or female. Let's assume it was female.

INDISPUTABLE CONCLUSION: **The first human was of a specific gender.**

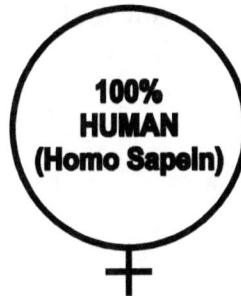

FACT: If evolution is true, a species that was not human gave
birth to the first human.

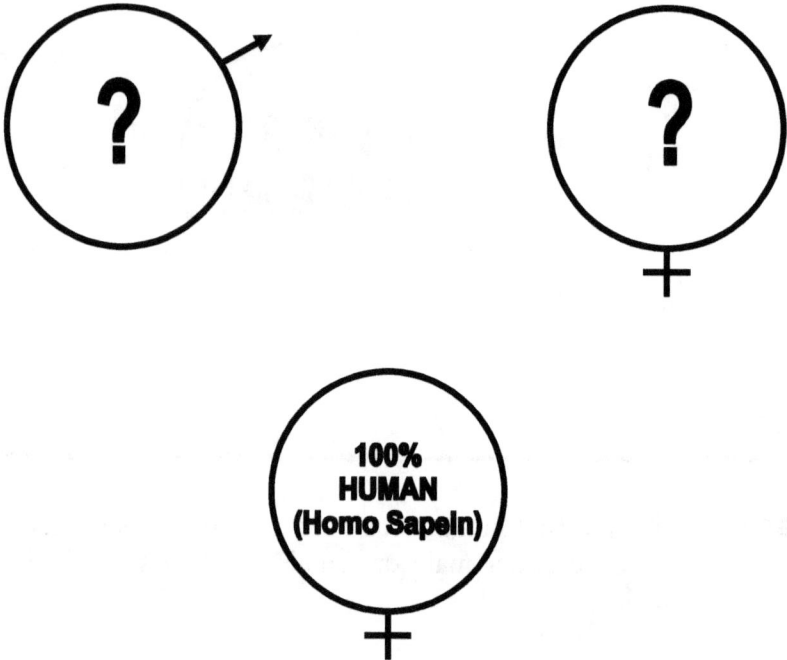

INDISPUTABLE   **The first human had two birth parents,**
CONCLUSION: **one male the other female**

FACT: Science claims that Humans and Chimpanzees share a common ancestor from six million years ago, which explains the 97% commonality in our DNA

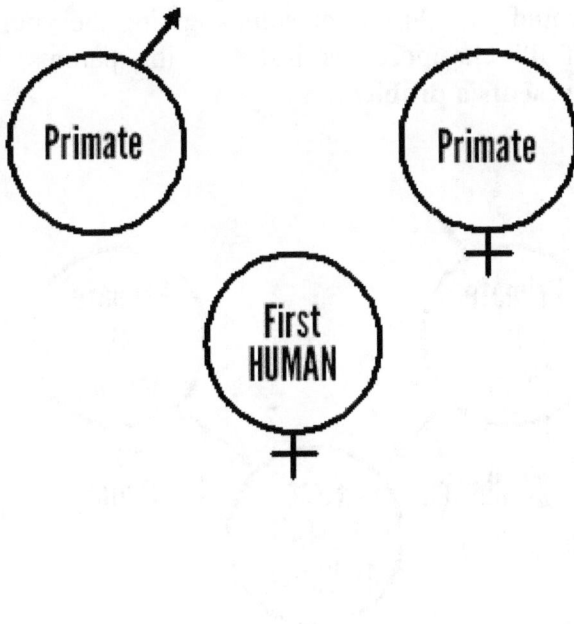

INDISPUTABLE CONCLUSION: If evolution is true, the first human's birth parents were primates

FACT: All physical traits are passed to the next generation through the chromosomes of the parents in sets of two. The chromosomes carry the genes, which determine all physical traits, as well as the sex of the offspring.

All primates have 48 chromosomes, so each parent would pass 24 chromosomes, giving the young a total of 48 chromosomes just like the parents. **But this presents a problem.**

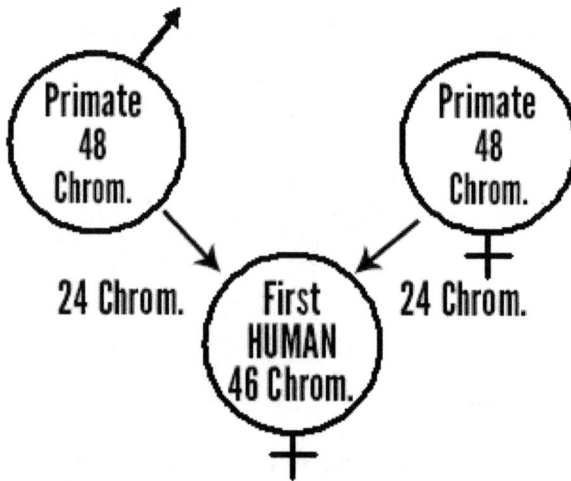

INDISPUTABLE CONCLUSION: **Humans have 46 chromosomes, not 48 chromosomes, _which proves we did not evolve from primates_**

The Theory of Chromosomal Fusion
This is science's latest attempt to patch The Theory of Evolution,
in order to make it appear feasible.

FACT: Scientists recently noticed a similarity between the human chromosome #2 and two of the Chimp's other chromosomes. They concluded that the first human was a result of a mutation, which fused two of early primate's chromosomes together and resulted in the human chromosome #2. By reducing the chromosome count by one set, or two chromosomes, it would bring the primate's chromosome count to 46, that of a human.

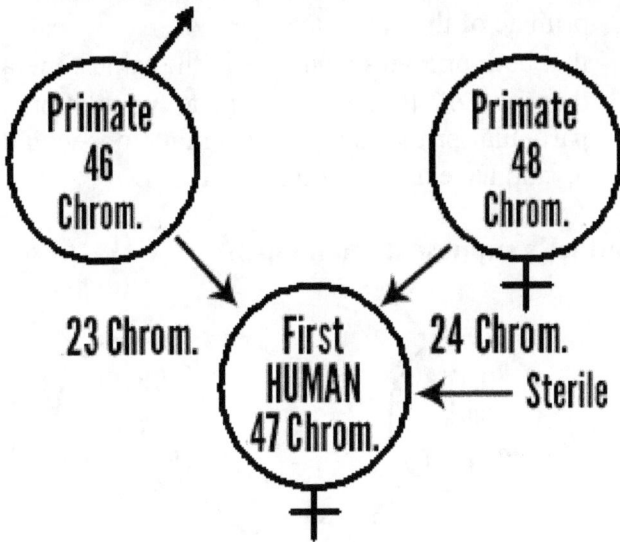

Primate
46
Chrom.

Primate
48
Chrom.

23 Chrom.

First
HUMAN
47 Chrom.

24 Chrom.

⟵ Sterile

INDISPUTABLE CONCLUSION: **Offspring who receive incomplete sets of chromosomes are born sterile and cannot reproduce.**
Proof: All mules are born sterile with 63 chromosomes (Horse 64, Donkey 62)

New Supposition for the Theory of Chromosomal Fusion

Suppose two primates with the exact same mutation
(46 chromosomes)produced an offspring.

FACT: **What are the odds...**
- that such a mutation is possible at all?
- that the affected primate would survive to adulthood? (mutations are abnormalities)
- that the identical mutation could happen to a second primate? (needed to produce an offspring)
- that the second exact mutation would happen to a primate of the opposite sex?
- that both primates would be in the same generation?
- that out of the entire primate population, these particular primates with exact mutations would meet up and have an offspring together?

But let's suppose it did happen.

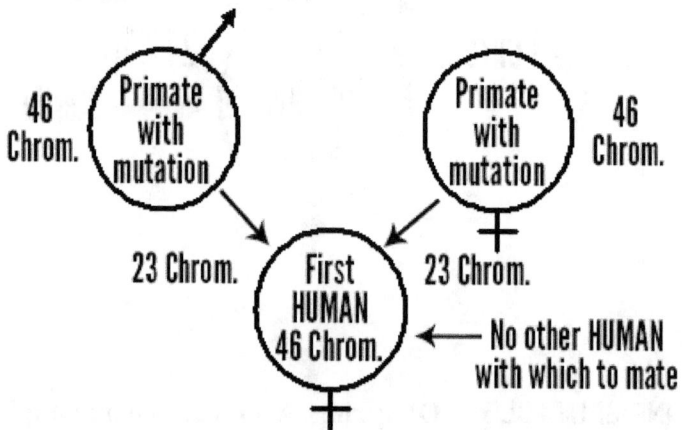

46 Chrom. — Primate with mutation

Primate with mutation — 46 Chrom. ♂

23 Chrom. — First HUMAN 46 Chrom. — 23 Chrom.

←— No other HUMAN with which to mate

♀

INDISPUTABLE CONCLUSION: **Theory Failure! Only like species will mate. Example: Even though all primates have 48 chromosomes, chimps will only mate with other chimps, orangutans with other orangutans.**

The Only Other Possibility

FACT: A very simple law of nature renders evolution impossible. Life requires both male and female to unite in coitus, to produce an offspring of like kind. This fact is stated sixteen times in the book of Genesis, "after its own kind."

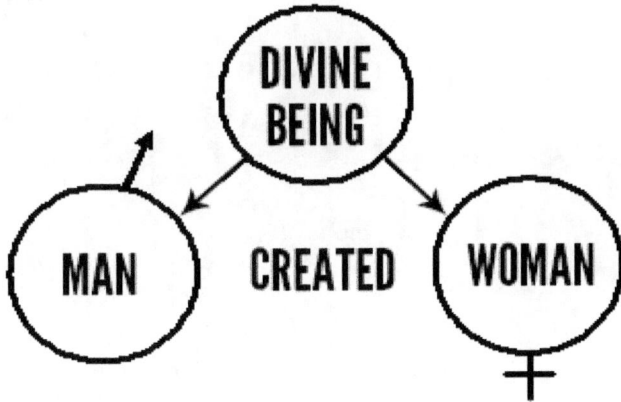

ADDITIONAL PROOF:

- Scientific evidence supports creationism, not evolution. Evidence shows that each plant and animal species appeared suddenly on earth, and remained constant until either becoming extinct or continuing to this day.

- The cells of all primates contain an acid known as Neu5gc, which the human immune system attacks when introduced. This is why the body always rejects transplanted tissue from primate to human.

- There are increasingly fewer species on earth. If evolution were true, earth would be teeming with new species.

- The common field mouse (a rodent created 100 million years ago) has more DNA in common with humans than Chimps do. Mouse – 99%, Chimp – 97%. DNA commonality among different species only proves a common Creator, not evolutionary relationship.

9 781602 641662